Praise for *Wait, How Do I Promote My Business?*

"Danny Rubin — one of the smartest, most enterprising professionals I know — offers compelling, crisply written, actionable advice on how to leverage strong writing skills to promote one's business. His professional mantra: 'Write well, open doors.' Well said."

— PATRICK FORD
WORLDWIDE VICE CHAIRMAN AND CHIEF CLIENT OFFICER
BURSON-MARSTELLER

"Danny has created the definitive how-to playbook for entrepreneurs who want to build their brand. It's everything they don't teach you in business school, but should. You need this on your bookshelf."

— MATT HAYES
HEAD OF MARKETING
LEESA SLEEP

"There are hundreds, if not thousands, of wisdom bombs in the book. Go ahead and read a few pages, take some notes and then buy this TODAY!"

— ZACK MILLER
FOUNDER
HATCH STARTUP INCUBATOR

"This book is a blueprint for getting on the offense and making something happen for your business. Business owners who are questioning their pitch and need some assistance streamlining can follow these tactics to succeed."

— HAMILTON PERKINS
FOUNDER
HAMILTON PERKINS COLLECTION

Oct 2017

Wait, How Do I Promote My Business?

100+ Attention-Grabbing Templates for Website Content, Press Releases, Crowdfunding & More

Danny Rubin

Library of Congress Control Number: 2017905456

ISBN: 978-0-9963499-4-9 (Print)
978-0-9963499-3-2 (Ebook)

Printed in the United States of America

Cover design: Paul McCarthy

Interior design: 1106 Design

TO ANYONE WHO ONLY KNOWS ONE SPEED: HUSTLE.
THIS BOOK IS FOR YOU.

Introduction

Inside most businesses lurks a widespread yet undiagnosed problem. It's not a flashy buzzword or slick motivational speech with millions of online views.

But each day, the issue is central to the success or failure of your operation.

Can you guess? **It's writing skills.**

With every email, website page, press release and grant application, the quality of your writing speaks volumes about your team and your work. The same goes for employee LinkedIn profiles, crowdfunding campaigns, networking messages, blog posts and more.

Hey, it's Danny Rubin. As I demonstrated in my first book, *Wait, How Do I Write This Email?*, a collection of 100+ email guides for job seekers, strong writing skills are critical when you need to find a job and grow into a career.

In *Wait, How Do I Promote My Business?*, I go one step further and demonstrate why you again need writing skills as you build a company/organization. I don't need to tell you the business world is a competitive place. Who wins out? The people who know how to command readers' attention and drive them to action.

In this book, I provide step-by-step instruction for common writing challenges in the business world. I rely on my years of experience writing copy for clients in a broad range of industries: schools, law firms, trade associations, startups, hospital systems, retailers, financial planners and more.

No matter the industry, the same writing rules apply: brevity, rich storytelling, authenticity and a curiosity in other people. In the pages that follow, I weave those concepts into everything from an email announcing your new business (page 76) to a blog post about a recent company success (page 175) to a "project description" on Kickstarter (page 198).

With each template, I walk you through the content from start to finish with notes and observations along the way.

Throughout the book, I also share stories of my own success and the writing strategies I used to make it happen.

All right, enough talk. I will end the introduction the same way I did in my first book.

Let's get started.

We have important work to do.

CONTENTS

Author Notes

Author notes

Before you begin, two points about the book:

1. I use the writing lessons and templates in the book for my corporate communications workshops. They are the foundation of everything I teach. While I know strong writing skills can boost your business, I don't guarantee my techniques will lead to new relationships, deals or revenue. I *can* promise you will capture the reader's attention, and that's a great place to start.
2. Every person, school, business and organization I name in the book is fictitious.

Chapter 1
Business Writing Basics

INTRODUCTION

My professional mantra is "Write well, open doors" — a philosophy that applies to every aspect of our careers.

When you need a job, it's critical to write networking emails and job applications with precision.

And once you're in a job or running your own business, writing is once again paramount. In our careers, we must always compete for peoples' attention, and how we write can either pull them in or send them away.

In my "Business Writing Basics," I divide the topics into two camps:

▸ How to be brief

▸ How to say it best

Each section contains practical writing and editing instruction. Every day on the job, you have opportunities to apply these lessons and become a better writer.

Think of writing like learning a language or musical instrument. If you hone the ability a little each day, you will see improvement. And with stronger writing comes more effective emails, reports, presentations, grant applications, social media posts, press releases...

...you get the idea. Everything improves and you go further in your career.

Up first: a short lesson on brevity.

How to be brief

WRITE LESS, SAY MORE — THE POWER OF BREVITY

There is a common misconception when it comes to writing that is professional in nature that a person must write in a verbose manner to come across as intelligent.

I am sorry. Let me do that again.

People often make a mistake in thinking that writing long-winded sentences with big words helps them appear smart.

Actually, let me try this one more time.

You don't need to write a lot or use big words to sound smart.

Now, that's better.

Too often, people write sentences like the one at the top when they should choose version #3. The main culprit, in my view, is the loathsome college essay. Only in college are we forced to write a paper a certain length. We develop strategies that balloon our paragraphs so we can fill out eight, 10 or 12 pages.

In the real world, most people don't enjoy reading lengthy emails, reports and presentations. It's extra work and burdensome. Worst of all, trying to write beyond our skill level screams, "I'm in over my head!" and doesn't help us stand out.

When you write with brevity, you make your points quickly and shrewdly. You don't waste words and, in doing so, don't waste a person's time. A vendor or client, for instance, then sees you as sharp and courteous.

The secret to brevity (and, in turn, clarity) is something we are rarely taught growing up:

Write like you are talking to a friend.

I don't mean write in internet jargon or shorthand. Whenever I am stuck on a sentence, I step back from the computer screen and ask myself, "OK, what am I trying to say here?" Rather than come up with the most eloquent way to make my point, I write it out in plain English as if talking to a buddy. And once I have my conversational sentence, then I go and attack it with a red pen.

Let's use the examples from the top.

The before:

There is a common misconception when it comes to writing that is professional in nature that a person must write in a verbose manner to come across as intelligent.

The after:

You don't need to write a lot or use big words to sound smart.

First ~~things first~~, I located the subject ("You") and led with it. Active voice feels more confident to the reader.

To write the shorter sentence (version #3), I sat up from my computer and asked, "What am I trying to say?'" I stopped trying to be clever ~~with it~~, and the words found their way onto the page.

I also have a habit of being ~~very~~ critical with the ~~number of~~ words I use in each sentence. Once I write something, I ~~go back and~~ decide if ~~each and~~ every word ~~I just wrote~~ deserves to be there. Say to yourself: if I remove this word, would the sentence still make sense? If I remove this sentence, would the paragraph make sense? And the ultimate: do I ~~really~~ need this paragraph?

Speed is key. When people read your business correspondence, you need to be ~~very~~ respectful of their time. Don't write five huge paragraphs ~~that go on and on~~. ~~Be tough on yourself and really give them just what they need to know~~. You are better off making one or two main points (or telling one great story) ~~rather~~ than trying to jam everything you know into someone's brain.

And when you finish editing your work, ~~go back and~~ edit again. After that, ~~go back and~~ edit some more. A client may never tell you he/she loved your email or report, but ones that are tightly written ~~and well-composed~~ will leave an impression.

Most of all, ~~y~~ou will stand out. College did not prepare us ~~very well~~ to write in the business world. But those who ~~take it upon themselves to~~ learn ~~to harness~~ the power of brevity will have an edge every time.

Editing basics for business emails

Every email or document deserves the same steps in the proofreading process.

Refer to the following checklist before you hit "Send" on important business-related emails.

▸ Is the main point of your email at or near the top? Make sure you don't bury the info in the middle of the message or at the end.

▸ Is each person's first and last name spelled correctly?

▸ Are company and product names spelled correctly?

▸ Do all of your links work?

▸ Are there people on this email who won't know what certain abbreviations or expressions mean? If so, spell out the abbreviation on first reference and clarify the expressions.

▸ Do you have any paragraphs with 4+ sentences? If so, break the paragraph into smaller sections.

▸ Once you have covered the list, print out the email and read it aloud to yourself. If everything looks and sounds good, you're ready to send!

Bottom line up front (BLUF)

Busy people need answers ASAP. They don't have time to search around for what you need or hope to convey.

That idea applies directly to business communication. As you draft emails on the job, you need to think "BLUF" every time.

BLUF is short for "bottom line up front," and it's an approach that allows the reader to spot your main point right away. Then, the rest of the email supports your central statement or argument.

Here's an example for an email in which someone pitches about a new company product.

Incorrect version

Subject line: Hoping to reconnect

Hi John,

I hope all is well.

As a reminder, I'm Jim Langford, vice president of product development at Acme Corporation. For the past nine years, we've been an industry leader in emergency medical equipment and now provide supplies to two dozen hospitals in the Pacific Northwest.

Here's a short video that chronicles our growth and our plans for the coming year. In fact, you might even recognize the place where we shot the video (the park right outside your offices).

And here's a big profile story we landed in The Daily News. We are proud of the coverage because it shows how far we've come.

I'm writing you now to tell you about our newest product, the Acme Device 2000. It's a smarter, more efficient way to manage patient data across multiple medical centers. I think your team will appreciate what we've created. Here's a quick breakdown.

Please let me know if you'd like to talk further about the Acme Device 2000. I'm happy to set up a phone call when it's convenient for you.

Thanks,

– Jim Langford

Why is this version "incorrect"? Simple. Where is the most important part of the email (info on the new product)? It appears in the fourth paragraph ("I'm writing you now..."). That's WAY too far down.

Remember: BLUF. Make sure the person reads the most important part of your email first. Then, the rest of your message bolsters your argument (in this case, that Acme Corporation makes great medical devices and is worth the reader's time).

Also, be sure the "main point" is clear in the subject line too.

Correct version (the "BLUF" is in bold)

Subject line: Information on Acme's new Device 2000

Hi John,

I hope all is well.

As a reminder, I'm Jim Langford, vice president of product development at Acme Corporation.

I'm writing to tell you about our newest product, the Acme Device 2000. It's a smarter, more efficient way to manage patient data across multiple medical centers. I think your team will appreciate what we've created. Here's a quick breakdown.

For the past nine years, we've been an industry leader in emergency medical equipment and now provide supplies to over two dozen hospitals in the Pacific Northwest.

Here's a short video that chronicles our growth and our plans for the coming year. In fact, you might even recognize the place where we shot the video (the park right outside your offices).

And here's a big profile story we landed in *The Daily News*. We are proud of the coverage because it shows how far we've come.

Please let me know if you'd like to talk further about the Acme Device 2000. I'm happy to set up a phone call when it's convenient for you.

Thanks,

– Jim Langford

As you compose emails, ask yourself:

- What's my BLUF? What's the most essential information?
- Do I have my "bottom line up front"?
  - If not, how can I rearrange the email so the main point comes first?

How to improve everything you write in three minutes

The tutorial below allows you to improve anything you write in a few short minutes. Keep these pages handy!

Step 1: When you finish your document, hit CTRL+F to bring up the search function.

Step 2: One by one, look for these words and delete/amend them.

- very, just and really (remove all three)
- that (delete, as in "I believe that you are correct")
- quite (delete, excess word)
- thing (replace with specific word for the "thing")
- utilize (switch to "use" or pick another verb)
- get or got (pick another, more descriptive verb)
- -ing verbs ("At the meeting, we will be discussing..." becomes "At the meeting, we will discuss")*

9

Step 3: Read over your work to check your edits.

*The "-ing" verbs bullet point deserves further explanation. At the start of an email, you may want to use the phrase "I am writing" as in "I am writing to introduce myself." In my view, that's an acceptable use of an "-ing" verb because it's the best way to begin. "I write to introduce myself" is too stilted.

In many other cases, you can cut the "-ing" and the sentence still makes sense. Here's one more example with multiple "-ing" words:

*At Acme Industries, we are all about **bringing new ideas to life and understanding the needs of the consumer.***

Let's chop down the two "-ing" verbs.

*At Acme Industries, we **bring new ideas to life and understand the needs of the consumer.***

The original sentence has 19 words.

The revised sentence has 16 words.

Brevity makes you sharper. Plain and simple.

Once you hit "Send," you can't get it back

I won't drag out this point because it's fairly obvious. But it still deserves a special discussion.

Once you press "Send" on a work email, you can't get it back. That's it. Think hard about the message and decide if every word is necessary. These are the moments when editing can save you and your reputation.

Yes, the information could come back to haunt you. What's worse, the poor choice of words could become part of a human resources dispute or even a document disclosed in a court case.

Every email you send adds to your paper trail.

Here are examples of emails you may want to reconsider:

- Speaking ill of a colleague, coworker, client or vendor
- Divulging sensitive information on your company or private information on a client/coworker
- Insulting your company's management

You can always "write an angry letter," walk away from the computer for an hour and then delete it altogether. Sometimes it's therapeutic to put your words on the page and find you don't want to send the message at all.

Why you need a #personalhashtag for business

In my first book, *Wait, How Do I Write This Email?*, I introduced the concept of a #personalhashtag as a smart way for job seekers to promote their career accomplishments on social media.

For this book, a collection of writing guides for working professionals, I believe the #personalhashtag applies again.

Here's what I wrote in book one (from page 51):

Let's say you send out dozens of resumes as you look in every direction for a job. Right below your name, you add a personal hashtag so the resume stands out (for an example, check out #dannyrubinportfolio). You also include the hashtag at the top of cover letters and in your email signature. Employers are likely to stop cold and check out your hashtag.

Why? Because they've never seen anything like it.
With a personal hashtag campaign, you share what an employer needs to see.

#whatasimpleidea

Now, we will extend the concept to the work world. How can you use a #personalhashtag to promote your career and the success of your own business (or the place where you work)? You can tweet or post on other social media platforms about content like the following:

- ▸ Recent blog posts from your company website
- ▸ News articles that mention or feature your company
- ▸ Awards you or your company won
- ▸ Business-oriented photos that showcase you and/or your company
- ▸ Testimonials from happy clients

Then, you give the world a quick snapshot of your highlights via social media. The person only needs to search your hashtag and voila — your best stuff is waiting for them.

In the first book, I use a closing line that also applies here.

What will you call your #personalhashtagcampaign? And what will you share? Here are some examples for Acme Corporation:

- ▸ #AcmePortfolio
- ▸ #AcmeGreatestHits
- ▸ #AcmeShowcase

In business today, you need to tell your story faster and smarter than the next person.

#getwhatsyours

How to craft an effective email signature

In our careers, we need to compose email signatures for four main scenarios:

- – Unemployed and looking for work
- – Freelancer
- – Working professional with a job
- – Working professional who also attends school

On the following pages, you will find templates for the different scenarios. Before you begin, heed these four rules:

- Remember: less is more. You don't need to give people nine ways to contact you. Focus on the best ones (ex: phone, email, Twitter and LinkedIn) and make it easy on you and them.

- Stay away from a signature that's one big image. Keep it as text so email services won't block people from seeing it.

- Make links long enough so they're easy to click on mobile devices.

- Include your #personalhashtag when appropriate, as we discuss on page 11.

- The hashtag lets you give people a quick look at your accomplishments or those of your business.

Unemployed and looking for work

Your Name
EMAIL | CELL: XXX-XXX-XXXX
TWITTER | LINKEDIN | #YOURNAMEPORTFOLIO
BLOG/PORTFOLIO

EXAMPLE:

Jane Doe
XXXX@_____.COM   CELL: 555-555-5555
TWITTER | LINKEDIN | #JANEDOEPORTFOLIO
JANEDOE

Explanation:

The above is a universal template for someone who needs to find a job. It includes relevant contact information, social media/URL and the #personalhashtag. Again, if you don't have all of these items, that's OK. But the links included are the *right* amount of info and won't overwhelm the reader.

Freelancer

Your Name

EMAIL | CELL: XXX-XXX-XXXX

TWITTER | LINKEDIN | #YOURNAMEPORTFOLIO

BLOG/PORTFOL O

_____

E X A M P L E :

Jane Doe

XXXX@_____.COM | CELL: 555-555-5555

TWITTER | LINKEDIN | #JANEDOEPORTFOLIO

JANEDOE

Explanation:

The email signature is, in my view, the same for a freelancer and someone who's unemployed You need to provide your contact info and social media/URL links. You can also consider a #personalhashtag. Give people a quick snapshot of your work and ability right from the signature.

Working professional with a job

Your Name

TITLE , COMPANY

EMAIL | CELL: XXX-XXX-XXXX (INCLUDE OFFICE LINE, IF NECESSARY)

TW TTER | LINKEDIN | #YOURNAMEPORTFOLIO

COMPANY URL

_____

E X A M P L E :

Jane Doe

REGIONAL MANAGER, ACME CORPORATION

XXXX@_____.COM | CELL: 555-555-5555

TWITTER | LINKEDIN | #JANEDOEPORTFOLIO

ACME CORPORATION

Explanation:

The job title appears below the name followed by similar contact information and social/URL links as previous email signatures. It's important to remember a company might require the email signature be done a certain way. What I have provided is a general, one-size-fits-all approach.

> *Note: Jane still includes her personal hashtag. Now, she produces work for Acme and can use the hashtag to promote recent company projects and successes. The hashtag #janedoeportfolio is an ever-evolving place to highlight her latest and greatest achievements — personal or those of her company.*

Working professional who also attends school

Your Name

TITLE , COMPANY

EMAIL | CELL: XXX-XXX-XXXX (INCLUDE OFFICE LINE, IF NECESSARY)

TWITTER | LINKEDIN | #YOURNAMEPORTFOLIO

COMPANY URL

[DEGREE] CANDIDATE, [DEGREE PROGRAM] [SEASON, YEAR] — COLLEGE OR UNIVERSITY

E X A M P L E :

Jane Doe

REGIONAL MANAGER, ACME CORPORATION

XXXX@_____.COM   CELL: 555-555-5555

TWITTER | LINKEDIN | #JANEDOEPORTFOLIO

ACME CORPORATION

MASTER'S CANDIDATE, PROJECT MANAGEMENT (SPRING 2021) — TECH UNIVERSITY

Explanation:

Jane decides to head back to school for a master's while she works and needs a signature to reflect both responsibilities. The easiest and cleanest way is to focus on her job first and add graduate information at the end.

How to say it best

HOW TO ADDRESS PEOPLE PROPERLY

With business emails, it's all about appropriateness. You would hate to address someone the wrong way and turn the person off before you even begin the message.

As I provided in *Wait, How Do I Write This Email?*, here's a chart based on your age and the age of the recipient.

To determine the recipient's age (or the person's general age range), consider looking at the person's LinkedIn profile for the year he/she graduated college.

NOTE: If you don't know someone's age, always take the safe route and use "Mr." or "Ms."

If you're ages 18–22

Age of Email Recipient	Nature of Relationship	Title in the Email
Under 35	Never met or an acquaintance	"Hi [first name]"
Under 35	Friendly or have spoken in person	"Hi [first name]"
Over 35	Never met or an acquaintance	"Hi Mr./Ms. _____" Use "Ms." unless you know the woman wants to be addressed as "Mrs."
Over 35	Friendly or have spoken in person	"Hi [first name]"

16

Deeper Insight

When you're 18–22 (and likely a college student), you need to address people formally if they're over age 35. That age (35) is my benchmark for when someone crosses over into REAL adulthood. Translation: they have departed young adulthood and are now senior business people.

> NOTE: Also, it's standard etiquette to address women as "Ms." versus "Mrs."

If you're ages 23-30

Age of Email Recipient	Nature of Relationship	Title in the Email
Under 40	Never met or an acquaintance	"Hi [first name]"
Under 40	Friendly or have spoken in person	"Hi [first name]"
Over 40	Never met or an acquaintance	"Hi Mr./Ms. _____" Use "Ms." unless you know the woman wants to be addressed as "Mrs."
Over 40	Friendly or have spoken in person	"Hi [first name]"

Deeper Insight

Similar rules for people ages 18–22, but I moved everything up five years since you're a bit older.

If you're ages 31–40

Age of Email Recipient	Nature of Relationship	Title in the Email
Under 40	Never met or an acquaintance	"Hi [first name]"
Under 40	Friendly or have spoken in person	"Hi [first name]"
Over 40	Never met or an acquaintance	"Hi [first name]"
Over 40	Friendly or have spoken in person	"Hi [first name]"

Deeper Insight

Once you enter your 30s, everyone is on a first-name basis. You are now a mature working professional and don't need to add a "Mr." or "Ms." when you address people in emails.

> *NOTE — Exceptions to the rule: dignitaries, public/elected officials and "important" people (ex: you email a CEO and ask if he/she will give a commencement speech for a graduation).*

If the person is a doctor, make sure to use "Dr." as a formal title.

Comma after "Hi"?

You will notice throughout the book that I often begin emails with:

Hi _____,

Technically, I should also put a comma after "Hi" so it would look like:

Hi, _____,

Looks weird, right? I know. The two commas throw me off.

Now maybe an English teacher would disagree, but this is where I "learn the rules and break them."

I think it's cleaner and less distracting to go with:

Hi _____,

Rules and structure matter, but so does readability. And you don't want to distract the email recipient right from the jump.

For me, it's "Hi" and then the person's name followed by a comma.

That's my story, and I'm sticking to it.

And finally...why do I use "Hi"? Again, it's a gut feeling on the best "intro" word for the majority of business situations.

The other options don't feel right.

▸ Hello: too flat and impersonal

▸ Hey: too comfortable

▸ Dear: too formal

▸ Hey Hey: don't even think about it

And if I don't have the person's name, I go with "Hi there," — it's safe, courteous and won't rub the person the wrong way.

How and when to follow up on common business emails

How many times have you sent an email and waited forever for a response?

All the time, right? Like every single day. You're anxious, want an answer and are tempted to send the "Did you see my email?" reply right away.

Not so fast. I included a chart (updated from my first book so it's more focused on business scenarios) to help us through the trickiest email situations.

Business Challenge	Amount of Time	What to Write	Additional Notes
Need an urgent response	Give the person 1-2 hours to respond. If no answer, it's time to check in.	"Hi there, I want to make sure you saw my email. Please let me know when you have a chance."	We all have smartphones today and see email immediately. If a person knows you need a response ASAP, then you have every right to follow up within 1-2 hours.
Wait on an update from a client	Wait 24 hours for the update. If it never comes, send another email.	"Hi there, Please let me know if I can expect the project update today."	Don't jump down the person's throat looking for the info. Let your request simmer, and after a day has passed check back in.
Wait for confirmation after you send over a proposal, presentation or other critical item	Lay low for 24 hours.	"Hi _____. My name is _____; and yesterday I sent over [type of item] about [topic at hand]. Please let me know you received it. Thanks very much."	24 hours is enough time to receive a "Thanks, we received it." If your follow-up email doesn't work, call the office and ask again — politely.
Wait on someone to network for you	Let 2-3 days go by.	"Hi, I want to make sure you can still send the networking email to _____ on my behalf. I appreciate you doing this for me."	Someone has offered to do you a favor. Great. So give it time, and let people fit the "free" help into their own schedules. Don't be too antsy.
Send an email about doing business with someone new	Let 2-3 days go by.	"I want to make sure you saw my email from the other day about exploring business opportunities. Please let me know if you have a few moments to discuss."	If no answer after email #2, see the scenario below.
Can't receive a response over email at all	After 2-3 days, it's time to get on the phone.	Phone script: "It's [your first and last name]. I'm calling to follow up on [task at hand] and thought it might be easier by phone."	Don't be passive and wait for a response by email. Be confident, pick up the phone and stay in control.

The power of a "wrinkle" in business

Communications is a game of managing expectations.

Strangers who receive our messages often assume an email will appear a certain way. We know, for instance, what a standard "cold call" email or thank-you note is supposed to look like.

And when the message goes above and beyond what we expect, it can leave a lasting impression.

The tactic is what I call a "wrinkle," a move that catches people by surprise and makes us memorable.

Here are two examples to drive home what I mean.

"Cold call" email

Readers assume the message will be impersonal and a typical pitch about a product/service. The expectation is "I have seen this email 1,000 times. I know what I have received."

That's when you apply the "wrinkle" method. Let's use a template from inside this book, "How to introduce yourself to a company for the first time." After you introduce yourself and your business, you do something different.

From page 78:

I'm writing you because I think [person's company; for instance, "the Acme Pet Supply team"] would like to learn more about [name of company; for instance, "Acme Pet Treats"]. [Why should the person care? Why does your business matter? For instance, "The product is healthy, safe and selling out routinely on our website."]

[Then, one sentence to show you studied the company's website. It's a powerful way to prove you didn't send the same email to 100 business owners. Be as specific as possible. For instance, "Also, I enjoyed the photos from the recent Bark Bark 5K race you sponsored. The chihuahua with the running shoes? Too good!"]

The reader expected my email to be a cookie-cutter, impersonal message. Not so fast. I included research on their business and even linked to a project I found notable.

Suddenly, the mood changed. The reader wanted to click "delete" and be done with me. But since I made the email customized and authentic, the hope is the person thinks harder about a response and perhaps starts a dialogue.

Thank-you note

Want to write a thank-you note the person never forgets? Again, it comes down to managing expectations.

Let's say you held a grand opening event and want to send handwritten notes to guests who attended. The typical route is to write something like:

"Thanks so much for attending our grand opening. You helped to make the day a success. We hope to see you again soon!"

Look, there's nothing wrong with the "typical" approach. But it's also forgettable because it's clear you sent the same message to everyone. Where's the personalization? Where's the "wrinkle"?

Here's an example from another template in the book, "How to thank someone for attending an event."

From page 94:

Thanks again for coming to the [name of event; for instance, "Hill Valley grand opening] this past week. I'm glad you were able to attend and [the reason you're glad; for instance, "see our new store firsthand"].

[Then, the special reference; for instance, "Plus, wasn't the catered food delicious? It was so funny how we 'fought' over the mini grilled cheese sandwich. Next time, it's mine!"]

[If you don't have a moment the two of you shared, be specific about another topic; for instance, "I know you're not the biggest fan of country, but I hoped you still enjoyed the twangy country singer who performed!"]

————

The person expected a quick, generic thank you. Instead, you reference a specific moment from the night that you both shared. It takes the message from "that's nice" to "wow, what a special note."

In every writing scenario, the goal is to be unforgettable. The "wrinkle" method will make it happen.

5 phrases that make you look weak

As you compose emails (and in conversation too), remove these five words/phrases from your vocabulary. They make you look weak.

1. Just

"I *just want to ask you…*"

"It'll *just take a minute…*"

"I'm *just saying…*"

Weak, weak, weak. "Just" is a little word with big implications. Each time we use "just," it suggests we're wasting someone's time. No, if you have something important to say, then say it.

Well, anyway…it's just a writing tip.

See how that sounds? Weak.

Moving on.

2. Sorry

Don't apologize all over the place. In most cases, you didn't do anything wrong. "Sorry" is more like "Sorry for bothering you" or "Sorry for taking up your time."

Of course, if you *did screw up, then yeah…say "Sorry."*

But if you have worthwhile information to send in an email or say aloud, then go for it. Respect yourself and the value you add to the conversation.

3. I'm not sure if you can, but…

Sigh, such an inferior tone. As if the other person is SO important and SO busy that you need to kneel down and beg for assistance.

How about, "Are you able to…"?

Stay on equal footing with the person across from you. You're no worse (or better). Eye to eye is the way to play it.

4. I hate to bother you, but…

Similar to #3, "I hate to bother you, but…" connotes the other person has all the power in the relationship. Even if you're an intern, new hire or several years junior to someone at the company, you have every right to stand proudly and say, "When you have a minute, I'd like your opinion on…"

And let me tell you, plenty of business execs can "suddenly" find 15 minutes in their jam-packed schedules if someone wants their opinion. Maybe even 30 minutes or an hour.

5. I hope that's OK.

Don't give up authority in the conversation — you have the same rights to the territory. Instead, go with "Thanks for the consideration" or "I appreciate the help."

———

Well, I *hope* you like my advice. If not, *sorry* for the trouble!

…said the author you don't respect.

Your words set the tone. Use them wisely.

How to write an "elevator pitch"

In today's rapid-fire pace of business, a tightly written "elevator pitch" about yourself or your business is key.

When I use the phrase "elevator pitch," I mean the time in which you would travel between floors 1 and 5 — not 1 and 50. You have a few precious seconds to convey what you're about.

Here is my "elevator pitch" in two sentences. You will also find these lines on the back cover of the book.

Danny Rubin teaches students and working professionals the power of strong communication skills.

As a public relations professional and former TV news reporter, Danny understands how to land valuable media coverage and capture the attention of your target market.

By the time I reach the 5th floor, the other person in the elevator would say, "Oh, OK. Now I see what you do."

The "elevator pitch" contains:

– One sentence on the basics of my work
– One sentence on why my work matters

The "what" and the "why" — the two pieces you need so people understand the full value of your work.

Where can your "elevator pitch" appear? A few ideas:

▸ Website home page

▸ Website "About" page

▸ Brochure, flyer or other marketing piece

▸ Text within a corporate or promotional video

▸ LinkedIn profile page or business page

▸ On the back cover of the book you will write one day :)

To create your own description, ask yourself:

▸ If I had 5–10 seconds to tell a perfect stranger what I do, what would I say?

▸ Do I use terminology the average person wouldn't understand? How can I simplify the language?

▸ How do I demonstrate value to the market in my explanation?

Then, press the "up" button and take that next elevator ride with confidence.

Additional guides for effective email writing

My first book, *Wait, How Do I Write This Email?*, contains over 50 pages of advice on how to write better emails.

In the book, you will find tips like:

▸ Why you need to be careful when you use acronyms

▸ How to "let the words breathe" in the email body

▸ When to use exclamation marks in job search/work emails

▸ Why you should be a name dropper

As a preview, here's a short section on pronoun usage.

Four sneaky words that diminish our work

Pronouns are a nuisance and in particular the four in bold: **this, that, these and those.**

Time and again, the words create confusion and water down your message.

Emails, job applications, presentations. All over the place.

Exception to every rule: Sometimes we use pronouns to tease people on purpose. Like an email newsletter subject line that reads, "You're not gonna believe THIS one!" Or even the title to my book: Wait, How Do I Write This Email?

Once the pronoun lures the person in, you then need to explain what "this" means. Got it?

OK, let's explore the use of pronouns in our careers. Here's an example on a cover letter.

The person starts a new paragraph with:

"One reason I did that is because I need the right skills to be competitive."

In the writer's mind, the sentence is fine. "That" refers to the decision to obtain a master's of sustainable design from Big State University.

Not so fast. Readers need constant guidance and a nondescript pronoun leads them astray. The sentence should be:

"One reason I obtained the master's degree is because I need the right skills to be competitive."

Get my drift? Below, I have an easy way to remove the four troublesome pronouns when appropriate.

Instructions

1. Hit CTRL+F and look for the pronouns: **this, that, these** and **those.**

2. If the pronoun represents a word or phrase, consider if you should delete and replace. Here are a few examples:

- "I wrote that to prove a point" becomes "I wrote the grammar lesson to prove a point."

- "Sally gave me this to say thanks" becomes "Sally gave me the present to say thanks."

- "The Millers need these for the vacation" becomes "The Millers need the house keys for the vacation."

- "John handed me those to be helpful" becomes "John handed me the hammer and nails to be helpful."

3. Check your edits for clarity and comprehension.

4. Voila! You made your edits like a pro.

My Journey: Part 1
That time I decided to start a personal blog

I believe a successful career is the combination of thousands and thousands of small decisions that add up over time.

But every now and then, a single decision becomes one of the biggest moments of your life.

For me, the big moment was when I started a personal blog.

At the time (July 2012), I wanted a place to stoke my passion (writing), share advice with my peers and, frankly, fill time in the day. I worked from home and had a pretty quiet lunch break.

To learn how to begin, I asked to meet with a blogger friend (template on page 71).

I called my blog News To Live By, and with each column I highlighted the career advice "hidden" in the headlines. What can the news teach us — good or bad — so we can be better at our jobs and in life?

I opened up a new blog post screen in Wordpress, began to type and from that day my career was never the same.

Over the past five years (I'm writing you now in 2017), the blog has been the catalyst for every important career decision. It helped to define my purpose (teach business communication skills), write books, hone my speaking/presentation style, build online courses and develop a business model that's uniquely mine.

Today, I have a more robust website (DannyHRubin.com), and my blog (now called THE TEMPLATE) is one piece of the pie. But the journey began as a simple, personal website for about $135 in year one.

My website is my home base, my foundation, my sacred space on the internet. I have poured my heart into that custom URL, and it's given me so much in return.

If you're searching for your purpose and how to "make your mark," my advice is simple.

Start a blog. Share what you love. Let the audience tell you what it values most. Then **refine refine refine** your approach until you deliver the greatest value possible to someone else's life.

Yes, the journey of a million little steps starts with a single action.

Little did I know my first step would lead to a life-changing project: my book.

Chapter 2
Public Relations 101

PUBLIC RELATIONS IS NOT ADVERTISING

Right out of the gate, I want to clear up a common misconception about public relations or "PR."

PR is not the same as advertising. This is how I describe it to people:

Advertising is when you spend money to promote your business (ex: commercials, billboards).

PR is when you find creative ways to have others spread the message for you (ex: article about your business in the newspaper).

And no, advertising with a media outlet (ex: TV news station) does not entitle you to a *free* news story about your company. In most circumstances, the newsroom and the sales department do not interact. If they did, they would have conflicts of interest all over the place as to what they should or shouldn't cover.

News: "There was a shooting last night outside Harry's Drug Store."

Sales: "Well, Harry's is one of our biggest advertisers. Let's leave the story alone."

Can't go there.

Make sense? Advertising and PR are separate concepts and should be treated as such.

Whether or not you spend money with the media to advertise, you have every right to pitch your event or announcement to the press. The editorial team will then judge the merits of the story and consider coverage.

And when you do send off a press release, understand that effective PR is not the following:

Cross fingers and hope media writes about your announcement/ attends your event

How do you actually land media coverage? It's a combination of strategic thinking and plain ol' determination. In the following pages, I explain how to:

- ▸ Craft a winning press release
- ▸ Send the release to relevant media
- ▸ Manage media at the event itself
- ▸ Do follow-ups with media after the event takes place

As I always do with email templates and other writing guides, I made the press release instructions step by step and easy to follow. Here we go…

How to build relationships with the press

Effective public relations means more than sending out a press release in all directions when you have a story to tell. It also comes down to building relationships with the press **before you have news to share.**

The best way to make friends with the media? Praise their work.

You can send a complimentary email to a journalist who covered your story or another story altogether. Either way, your extra effort will help to build a relationship and maybe lead to PR opportunities down the road.

Here's a short story that happened to me as a PR professional.

My team staged a press conference for a nonprofit client. It was a home build for needy families. Several media outlets came, including the NBC affiliate news station in our market.

That night, the NBC reporter put her story on the air, and I thought she did a nice job. I had asked her to please put a graphic on the screen with additional information on how to apply for the homes, and she did just that.

The next day, I sent her an email to say thanks. I removed the client's name here for privacy reasons.

Hi Laura,

Thanks again for the coverage of the story on Wednesday about [name of my client]. The [name of my client] team said the phone's been ringing off the hook with people applying for homes so all the awareness worked.

Also, **thank you** for running the full-screen of the criteria to apply for the homes. The application info is critical so the community has all the facts.

Have a great day,

– Danny

[and then my email signature]

———

I hoped Laura would respond to my email with a "Thanks a lot!" Instead, within 30 minutes, my cell phone rang. A number I didn't recognize.

Me: "Hello, this is Danny."

Laura: "Hi, Danny. It's Laura from the NBC station. I wanted to call and thank you for your nice email. That really means a lot to me."

Me: "Oh, great. Glad you got my message. I was happy to send it. Thanks again for your help yesterday."

Laura: "Absolutely. Hey, I was wondering…today I'm working on a story about the financial fallout in Europe, and I need to interview a local financial planner. Do you have anyone in mind?"

Me: "Yes, definitely. Let me email you his name and number."

Laura: "Great, thanks so much."

OK, I'll stop there because I want to explain what took place in a matter of 30 minutes.

1. I sent a thank-you email to the journalist.
2. The journalist responded by phone to say thanks and then asked if I could help her on another story.
3. In a half hour, I built a strong enough relationship with the reporter that she felt comfortable asking me to assist her on a different story — all from a simple thank-you note.

If I had not sent the email, it's unlikely she would have called to ask for my help. I became top of mind.

Yes, in this scenario I worked in PR and had clients in different industries (ex: financial planner). For you, as a business owner, the "praise" email will go a long way too. Now, the journalist may think of you for other stories related to your business or industry.

Journalists love to have topic experts and go-to sources. It makes life easier when the news director gives them three hours to put an entire story together and have it live on the air. As a former TV reporter, I can tell you: it's a mad scramble.

A special note will make you memorable and hopefully put you on a journalist's "short list" of sources he/she likes to call — especially when in a bind.

And for you, a business owner in need of positive PR, a short list is a great place to be.

Why you need to join industry associations

It's tough to grow a business in a silo. By that I mean operating without networking and information exchange with other people in your industry.

I believe 100% in joining associations relevant to your business. They can open doors, lead to new relationships and give you access unavailable to non-members.

As I grow my own business as an author and speaker on business communication skills, I have joined various associations in the book publishing and career development spaces. Here's why:

▶ **E-newsletters and printed magazines:** Most of my memberships provide an e-newsletter and printed magazine (monthly or bi-monthly) full of useful content. Both publications continue to teach me more about book publishing and career development. They also introduce me to thought leaders and people I may want to partner with on certain projects.

▶ **Cheaper rates at conferences:** Association gatherings and conferences are terrific places to meet people and learn new skills. Usually, members receive discounted rates to attend.

▶ **Speaking opportunities:** If you want to find places to speak, it's often easier to approach an association where you're already a member. You're "one of us" rather than an outsider. Also, some associations have a "speakers bureau" where you can be featured as someone others can book for upcoming events. But to add your name to the list, you need to be a member.

▶ **Access to webinars:** Most associations have a library of webinars on critical industry topics and, as a member, you can watch for free or at a reduced cost.

▶ **Discounts on products/services:** Membership often grants you exclusive deals on products/services integral to your business. Why not save on the stuff you have to spend on anyway?

▶ **Inexpensive:** Most of the associations I joined have inexpensive annual dues for individuals (under $100). Yes, the cost could reach into the hundreds of dollars for a company to join, but it's well worth the investment for the other reasons I provide in this list.

▶ **Community of like-minded people:** Once you become a member, you're connected with people across the state, country or globe who love the same work you do. And that commonality can lead to interesting conversations, relationships, partnerships and whatever else.

▶ **Support system:** No one finds success alone. You will develop faster as a professional and go further as a team if you're attuned to your industry.

If you need a place to start, google "[name of your industry] association." The right group should pop up. Then, you're only a few clicks away from being "in."

Membership, as they say, *does have its privileges*.

How to write a press release

GENERAL OUTLINE

If you want the media to know about a recent accomplishment or upcoming event for your business/organization, the first step is a press release. You may have read that press releases are no longer in style and PR pros favor more creative digital methods. I maintain that, when done correctly, a press release is still the smartest way to share your news with the media. In the following chapter, I explain how to write each section of the release and make the biggest impact on the audience (ex: bloggers, news producers and news reporters).

The most important point about a press release: you must explain the "why" as soon as possible. That means you need to describe high up in the release why your announcement matters to the general public or a particular industry if you're being more targeted.

General information about a press release:

▸ Company contact info and media contact info must appear at the top.

▸ A headline and subheadline combination are ideal so you include the most pertinent information possible at the beginning.

▸ Do your best to keep the release the size of a one-page Word document — even if you paste the release into an email. One page worth of information is enough. Anything more and you'll likely lose the reader's attention.

▸ Provide links to websites, PDFs, etc…when possible but don't overdo it with links or hashtag campaigns. In general, three to four links (and one hashtag) are plenty for a single release. If you have too many, the reader won't know which ones are most critical.

▶ Write in the third person about your company. For example, **"Acme Corporation is excited to announce"** and not **"We are excited to announce."** For the release to feel professional, you can't use "I," "us" and "we."

Now let's build the press release step by step. Up first, the header information.

Header information

To give a press release proper structure, the top portion should include your company/organization logo in the upper left and your business address in the upper right.

Below the logo and contact info, it's customary to put the point of contact. The person may either be someone at your company or a public relations/marketing professional. Typically, you provide email address and phone number(s) for the point of contact.

If the person is comfortable giving his/her cell number, include it. We all know it's easier to track someone down by cell than through an office line. Let's say the media jumps on your release and would like to provide coverage. Well, you want the reporter to reach the point of contact ASAP. If the person is difficult to find, the reporter may move to another story. MAKE THE PROCESS SEAMLESS.

Here's what the header and contact information area can look like.

COMPANY LOGO

COMPANY NAME
123 MAIN STREET
ANYWHERE, USA 12345

MEDIA CONTACT:

JANE DOE, ACME PR
JANE@ACMEPR.COM
OFFICE: 555-555-5555
CELL: 555-555-5555

Headline and subheadline

The headline and subheadline are the first place to explain to the reader what's going on and why the story matters. The headline is the lead and the subheadline provides backup details to enhance the "pitch."

Typically, press releases come in two varieties: big announcement or upcoming event. Let's walk through a headline/subheadline for two fictional scenarios.

Big announcement

Acme Corporation Becomes First Company in Florida to Offer Student Loan Buyback Program

Decision could impact as many as 200 employees,
spur other companies to do the same

The headline tells the reader why the news matters. A company has made a bold decision about the student loan debt of its employees. Then, the subheadline gives more context around the "big news." In this case, we learn the student loan program could impact up to 200 people and may encourage other companies to launch similar programs. If all we ever read was the headline/subheadline, we would understand the main idea of the "big announcement."

Notice how the headline is large and bold while the subheadline is the size of the press release message itself and not in bold but rather italics.

Upcoming event

Acme Corporation to Kick Off 'First Friday' Happy Hours in Downtown Sacramento

Celebration will take place on June 2 from 5–8 p.m.
and feature live music, carnival games

With the event headline/subheadline, lay out the basic info in the first line and then provide additional info in the second line (ex: day/time and event details).

Again, give the reader the most pertinent info in the headline/subheadline, especially the day/time so a reporter/producer doesn't need to search all over for it.

Body of the press release

OK, we've written our header information and headline/subheadline. Now comes the meat of the release: the body itself.

Let's use the "student loan buyback" announcement from the headline/subheadline section to craft the body of the release.

Many PR veterans will tell you to open the press release with a standard, traditional line. Something like, "Acme Corporation has announced a decision to offer a student loan buyback program..."

But that's a snoozer of a sentence. So formal and rigid, right? I don't advocate a total creative writing exercise here; you need to be professional and polished. But you can still catch the reader off-guard with something a bit different.

———

ORLANDO, FL (October 12, 2016) — With student debt in the tens of millions for recent graduates in Florida, Acme Corporation has unveiled a new way to help young professionals pay off their loans. The Acme Student Loan Buyback Program will help to alleviate student debt for employees who work at least three years for the company.

Acme is prepared to spend $4 million over five years, which will help upwards of 200 employees with their student loans. Company executives feel they need to help younger employees strengthen their finances so they can make milestone purchases like homes and cars. The program is also a way to reward employees who make a longer-term commitment to the company.

"We are excited to roll out the Acme Student Loan Buyback Program and hope our employees are receptive," said Mark Wilson, vice president of human resources. "Student debt is holding back so many of our employees and, in turn, it hurts our company's chances to grow. We think the decision is a win-win."

The buyback program will begin in March 2017. Employees can apply to the program now as long as they will have worked at Acme for at least three years by March 2017.

Media: Acme executives will stage a formal press conference to announce the decision on Wednesday, October 19 at 11 a.m. ET at our corporate offices (123 Main Street, Anywhere, USA). Please arrive no later than 10:30 a.m. ET to ensure you capture the entire announcement. You can also watch the press conference online at this link.

###

Deeper Insight

There are several points I want to make about the press release copy.

- The intro line is still traditional and "newsy," but it's more engaging than "Acme Corporation has announced…"
  - Remember, you should treat a press release like a news article. That means it should not feature opinion. See how I wrote "…to help young professionals pay off their loans." I did NOT write "…to help young professionals pay off their **backbreaking** loans." "Backbreaking" is opinion. That's because, for some people, student loans may not be as difficult to pay off. Use quotes in the release to inject opinion — the rest of the text should be the facts.

- Lean on numbers to make your case. In the second paragraph, I use "$4 million," "five years" and "200 employees." Those stats give size and scope to the project so the reader thinks, "OK, now I know how big this news really is."

- In the third section, I include a quote. HERE is where you add opinion because it's a person talking and not the company itself. That's why it's OK for the person to use phrases like "Student debt **is holding back** many of our employees." It's what he, as a company official, thinks.

- At the bottom, I direct a message in bold to the media. If there's a "call to action" for the media (ex: attend an event or press conference), put the information in bold so they don't miss it.

- The ### at the bottom is the standard way to say "This is the end of the press release."

- ▸ Notice I have links in the release, but I don't overdo it. In this example, I have three:
  - ○ Link about the student loan buyback program
  - ○ Link for people to apply to the buyback program
  - ○ Link for the media to watch the press conference online
    - • Don't toss in too many links because then the media won't know which ones to click on. Same goes with a hashtag campaign, Twitter handles and other social media messaging. Decide internally the keywords and social content the media needs to know and then leave the rest out. Less is always more.

- ▸ At the bottom of the press release, you may also want to add a paragraph of background information on your company. Also known as "boilerplate" details, the paragraphs gives the media additional context on the business.

Look for news relevance

Need a topic to pitch to the press? Look to the news!

If you don't have an announcement or event to promote, think about how you can leverage the knowledge of your team members in a way that aligns with the news cycle.

For example, let's imagine Acme launched its student loan buyback program, and it's working well. Six months after the program began, Congress announces new legislation to forgive student loans for certain government workers. All of a sudden, reporters are scrambling to cover the story at the national and local levels.

That means the media in your market needs to find "experts" ASAP who can talk about student loan buyback programs. Lucky for Acme, it has VP of Human Resources Mark Wilson (remember him from our press release?) who now has considerable experience managing a student loan buyback program.

You can then write a press release to pitch Mark as an expert who is available for an interview. That way, if the media needs an expert on the topic, they can consider reaching out to Mark.

Here's the email template.

Subject line: Expert available to speak about student loan buyback programs

Hi there,

You may have seen the news about the Congressional legislation which forgives student loans for select government workers after five years of employment. The decision has led to increased discussion on student debt and the way it impacts young professionals in various industry sectors.

Mark Wilson, vice president of human resources for Acme Corporation, is available to speak about the potential impacts of the new legislation. Since October 2016, Wilson has spearheaded a program at Acme to buy back student loans for 175 employees. Wilson can talk about the positive outcomes from Acme's program and the challenges his team has faced along the way.

Please contact me, and I can schedule a time for you to talk with Wilson. I am available by email or cell (555-555-5555).

Thanks,

– Your first name

Email signature

Deeper Insight

Open the email with the news story at hand so you provide context. Then, offer your employee as the "expert" and explain why he/she is the right person to talk about the issue (ex: spearheaded a program…for 175 employees).

I recommend you provide your cell phone number. Reporters work fast so make it easy for them to reach you. If a reporter tries to call an office line and you don't answer, he/she will call someone else. You lose the PR opportunity. The competition wins.

Understand editorial calendars

You might be thinking: how do I land my company in the news?

Sometimes the answer sits on the calendar. An editorial calendar, that is.

Most newspaper and magazine editorial teams create calendars to guide content decisions throughout the year.

For example, an arts/style magazine in Virginia dubbed February/March 2016 its "Education" issue. In that edition, the magazine awarded Virginia's top teachers. If I had an education-related piece of news or information, perhaps the magazine would be receptive for the February/March issue.

Keep in mind monthly publications often have earlier due dates for story ideas. The deadline for the "Education" issue was December 31.

That's why it's important to keep a close watch on content calendars and submit ideas editors **need** in a given month. Once the publication decides "May will be our 'going green' month," its writers and contributors are on the hunt for stories on the environment and healthy living. And if your company or event fits the theme, you stand a far better chance at coverage.

Google the names of well-known local publications (magazines and newspapers, specifically) along with "editorial calendar." You can also look for a company's media kit. That's usually the best way to find the right info.

How to send a press release

AS AN ATTACHMENT? IN THE EMAIL BODY?

Many PR practitioners might disagree, but in my view the best way to pitch a press release is in the email and as an attachment.

In other words, you paste the press release content in the message area and attach the PDF.

That way, you give recipients two ways to view the message. People can quickly scan the pitch in the email and, if interested, they can download it too.

41

A huge part of pitching the press is to answer questions before the media needs to ask.

If you had not attached the press release, the journalist might need to write back with, "Can you send me the press release as a PDF?"

Send the release "both ways" and you're one step ahead of the game.

Email press release to a single reporter

With your press release ready to go, now you want to send it to the media. If there's a specific reporter/blogger you hope to target, send an email using the template below.

Subject line: [One line that sums up the news; for instance, "Acme Corporation to unveil Student Loan Buyback Program"]

Hi [reporter's first name],

I'm [first and last name], and I am a/an [job title] at [name of company].

Below I have included a press release about [the news at hand; for instance, "the decision by Acme Corporation to begin a student loan buyback program"]. [Then, one more sentence about WHY the news matters; for instance, "The initiative will help up to 200 employees — many of them millennials — so they can attain sound financial footing and make milestone purchases like a new car or home."]

[Then, a call to action; for instance, "Please let me know if you have an interest in the story. I can provide you with key Acme employees for interviews including Mark Wilson, Acme vice president of human resources."]

Thanks, and I hope to hear from you.

– Your first and last name

Email signature

Deeper Insight

In the email body, you succinctly explain the story and why it has value. Then, you tell the reporter/blogger you can connect him/her with people on your team for interviews (and call them out by name/title). Finally, you paste the release at the bottom of the email and attach it as a PDF to give the media two ways to view it.

Email press release to multiple reporters at once

If you want to send the press release to a group of reporters at once, you need to change the intro to the email, but the bulk of it stays the same.

The key differences:

▸ Send the email to yourself and BCC all recipients so no one knows who else received it.

▸ Open with "Hi there," rather than directing the email to a single person.

▸ If you send a release to media in a single market (ex: Tampa), include a line about why the story matters to people in that area. For instance, "Even though Acme Corporation is based in Orlando, the student loan buyback initiative could encourage similar efforts in other Florida cities like Tampa."
   ○ If you write out the city ("Tampa"), you will catch the reader's attention since you discuss the place where the reporters/bloggers live and work.

Who should receive the press release?

It's important to build a media list before you send out the press release. You may even want to set up a table to keep track of media and the responses you receive.

Something like:

Media Outlet	Contact Person	Phone	Email	Status
Acme TV	John Doe	555-555-5555	Johndoe@acmeTV.com	Received the release, deciding if they can cover event
Acme Radio	Jane Doe	555-555-5555	Janedoe@acmeradio.com	Will run a mention of event on air
Acme Website	Bob Smith	555-555-5555	Bobsmith@acmewebsite.com	Has not responded yet; need to follow up

To get started, here are different types of media outlets to consider for a local pitch:

▸ Daily newspapers

▸ Daily/weekly/monthly business journals and magazines

▸ Niche blogs and magazines (ex: lifestyle, family, health, tech)

▸ Radio stations that cover local news

▸ Local TV stations (to include the affiliates for NBC, ABC, CBS and FOX)

▸ Community access TV stations (often managed by city governments)

▸ E-newsletters that discuss local news/events (often administered by daily newspapers and business journals)

For event announcements, also make sure you post event details on calendars through websites like:

▸ Daily newspapers

▸ Lifestyle blogs

▸ Local TV stations

If you want to pitch a story throughout your state or nationally, it can be cumbersome to create a database of relevant media in multiple markets.

To save time, consider media list-building software and/or a press release distribution service. List-building software allows you to gather large sets of media contacts in a fraction of the time it would take if you did the research manually.

A press release distribution service will disseminate a press release to national/industry media so potentially hundreds of media outlets receive your release at the same time.

Each service costs money but will save you a lot of time with research and media distribution. In many cases, it's well worth the investment.

How assignment editors operate

When you email a press release to a TV news station (ex: NBC or ABC affiliate), it generally goes to an email account like news@acmetv.com. That means EVERY release goes to the same email address, and the inbox is a massive list of events and announcements.

The person who oversees the press release inbox is usually the assignment editor (although increasingly newsrooms are doing away with that role so multiple producers may see the release at the same time).

In the case of an assignment editor, the person acts like a quarterback and oversees what each reporter covers that day. As well, the person usually answers the phone when you call the newsroom.

In effect, the assignment editor is the gatekeeper for the newsroom. Since news stations operate morning, noon and night, the exact assignment editor will change over the course of the day. In most cases, one person handles the morning/early afternoon and then another person comes in for the late afternoon/night (through the 11 p.m. broadcast).

It's important to introduce yourself to the assignment editor and be personable. You will likely need that person to be an ally when it comes time for the newsroom to decide what it will cover on a given day.

Here's why:

Every morning and afternoon, most news teams gather for an editorial meeting. The assignment editor will either lead the meeting or be the person who tells everyone else (reporters, producers) about the best press releases in the stack (and it's always a big stack).

If you have won the attention of the assignment editor, he/she may highlight your event in the room. Moreover, he/she may *fight for your story and convince the room why it merits coverage.*

Bottom line: assignment editors are perhaps the most valuable people in the newsroom when it comes to your press release. Treat them kindly, and it could pay off.

How and when to follow up on a press release

Ah, the art of the media follow-up request. A delicate dance to be sure, but to stand the best chance of media coverage, you must check in with the media outlet in advance of the event.

Here's my approach with TV news stations:

STEP ONE: Let's say I have an event on October 22 where I want media coverage. A week out from the event, I send the press release to all relevant media. I either target individual reporters (see page 42) or send a BCC email to a group of reporters (see page 43).

I don't expect to hear back from the media outlet; each one receives A LOT of press releases on a given day. That's why I refer to the table I created (see page 44) and begin follow-up phone calls with each news outlet. Look for the phone number for the "Newsroom."

Here's a sample script you can use on the phone:

"Hi, my name is [your first and last name] with [name of company/organization]. Earlier today, I sent over a press release about [one sentence on what your release is about]. Can you confirm you received the press release?"

Time and again, the person on the other end of the phone tells me, "Actually, I don't think we got it." (That's code for "You probably sent it, but we can't find it.")

Then you say, "No problem. I can send the release over again. What email should I send it to?"

OR…the person could say, "Yep! We got the release. Thanks for sending."

Either way, use the time on the phone to encourage media coverage and stress how the story will have great visuals (if that's true). Reporters want to cover events that have the potential for compelling images. If your event will have them, then say so.

With this follow-up call, your goal is to make sure the media outlet has your release. This is not the time to make sure they can attend.

STEP TWO: When your event is two or three days away, it's time to follow up with the media once again. You can either check in with a reporter who said he/she would come or call the newsroom. This time, since you know the media outlet has the press release, your goal now is to push and see if they can cover the event.

Sample phone call script:

"Hi, my name is [your first and last name] with [name of company/organization]. Last week I had sent your team a press release about [one sentence on what your release is about]. I want to see if your team has plans to attend. Are you able to help me with that?"

The person on the phone might respond one of two ways:

1. "We haven't made decisions about what we're covering on that day yet."
2. "Your event is on our list, and we'd like to attend but can't make any promises."

Either way, you now have the event on their radar. And that's all you can ask at this point because in general, media outlets will make a final decision the morning of your event.

STEP THREE: The day of your event (in the morning), you need to call once more and see if the media outlet plans to attend. The best time to call is between 9:30 and 10:15 a.m. That's because most newsrooms have morning editorial meetings where they assign reporters and photographers (videographers) to various stories. Those meetings usually wrap up sometime after 9:30 a.m. So when you call the newsroom, the person who answers (usually the assignment editor), will have a plan for the day.

If you're told a reporter will cover your event, you can ask for the reporter's contact info (email, cell number) so you can contact that person if he/she needs any updated info (ex: location change).

Deeper Insight

The biggest takeaway from the three-step approach: it often takes a lot of legwork to bring the media to your event. It's not enough to send a press release, sit back and watch the media roll in. Everyone wants reporters to cover their events. You need to be persistent but in a polite, appropriate way. The instructions I provide allow you to strike that balance.

How to manage PR at your event

HOW TO WORK WITH THE MEDIA THAT ATTEND

At the event itself, whether it's a big announcement or some other company activity, you should provide assistance to the media while they are at your company or the location where you asked them to be.

There are some key pieces of information you want to have ready when the media arrive.

First, you need several copies of the press release on hand. That's because it's possible that a journalist has come to your event to capture information but has little to no knowledge of what's going on. The editor or producer may have said, "Run out the door and go cover this event. You'll learn more once you get there."

That means the person who arrives may have no context or background. Your job is to educate the journalist on the event and why it matters to the broader public.

You may also need to provide a few pointers on what the person should photograph or capture on video. As well, be prepared to explain who the journalist should interview. If you want a particular person to be quoted or go on camera, bring that person over to the journalist. In other words, take control of the moment.

You can also offer to show the journalist where to find the best footage, which is also known as "B-roll." So often, members of the press don't know what they need to know. And once they are on your territory, you can feel free to point out what matters most.

Ultimately, editors and producers decide what goes on the air, online or into a print publication. Most likely, though, they will use the content the journalist gathered at your event so it's incumbent upon you to make sure they obtain the best content possible.

How to send media your own photos/video afterwards

If media did not attend your press event — or some outlets did and others couldn't make it — you need to send your own photos/video to the newsrooms the same day via a filesharing site. For help with photo captions, go to page 159. That way, the reporters and producers can still make use of your content on the air, in print or online.

Don't send the media a giant list of pictures. Choose the best five to seven photos and leave the rest alone. Odds are, the media will use, at most, one to two photos.

Once everything is ready, grab the link for anyone to view the files.

If the email goes to multiple people, you can send it to yourself and BCC everyone. If you don't know the correct email address, call the media outlet and ask where they want files sent. It's usually something like news@acmeTV.com.

Here's the email template:

Subject line: Photos/video from [name of your event; for instance, "press conference about Acme student loan buyback program"]

Hi there,

I have provided photos/video [or, if you only have photos, that's OK too] from our event today about [remind the media about event details; for instance, "the announcement of Acme's student loan buyback program"].

<u>Here's the link to view the photos and video.</u> I have also provided captions for the photos that provide context.

Finally, I attached the press release for full event information.

Thanks, and please let me know if I can answer any questions.

– Your first name

Email signature

Deeper Insight

Make sure the link to the file-sharing site is viewable to anyone. Otherwise, media will respond to your email with, "I can't open it!" You'd rather get it right the first time.

Even though you need to send off photos/video in a timely manner, don't rush the process. Review the photos with your team and decide which ones the media should see.

How to thank a reporter/blogger for the coverage

If a reporter did a story about your event/announcement, it's always smart to send an email to thank him/her for the coverage. Why is it the right move? Let me count the ways:

- ▸ If the person did a great job, he/she should be commended. You may even want to copy the person's boss on the email so newsroom leadership knows about the quality effort.

- ▸ Effective PR means, in part, building relationships with the media. If you show reporters you respect their work, you will create a rapport.

- ▸ Once you have a relationship, you may then feel comfortable pitching stories directly to the reporter(s) down the road. That way, you have someone's undivided attention rather than sending the press release to the assignment desk (as we discuss on page 45) where it's tossed into a pile with all the other releases.

▸ Reporters rarely receive thank-you notes from viewers or readers. It's even less likely for a PR person (or someone at your company/organization who handles communication) to send a thank-you note. The reporter will never forget the note (and probably save it).

Here's the email template:

Subject line: Great job on [name of the news story; for instance, "Acme student loan buyback story"]

[Reporter's first name],

Thanks again for the terrific reporting about [name of the story; for instance, "the Acme student loan buyback program"]. You captured the announcement perfectly, and it's great to have exposure for what we're trying to accomplish.

In particular, I like how you [give a specific example of a moment in the story that you appreciate; for instance, "described our millennial employee, Robert, and demonstrated how much he needs help with his student loans. Thanks for taking time to visit his office even though it's apart from our main building"].

Again, nice work on the story. If it's OK, I'd like to keep you in the loop on future Acme announcements. Please let me know.

Have a good one,

– Your first name

Email signature

Deeper Insight

The most important part of the email is when you reference a specific example from the news story. If it's a print story, copy and paste the line you like into the email itself. If it's a video news story, recount a moment that stood out to you. When you include an actual example, you take the thank-you note from good to great.

And at the end, ask if you can send future story pitches. Hopefully, the reporter will say "Sure!" It's tough to find story leads all the time, and most journalists welcome the chance to have fresh ideas come their way.

Additional media outreach templates

NEW BUSINESS HIRES AND ANNOUNCEMENTS

Business announcements are a simple way to gain positive public relations for your company/organization. When you hire a new employee or promote someone, send the announcement to the appropriate media outlet. Usually, it's your local newspaper and/or a business newspaper/magazine that covers your market.

The best way to locate the right publication is to google "[your city] business journal." The top one to two results will generally be the correct media outlet(s) to submit your business announcement. A few other points before I provide template options:

▸ The publication may charge for the business announcement. If the announcement runs in a print edition, the publication often charges per line. For example, where I live in Virginia the local newspaper makes you pay to run business announcements in print. The weekly business journal (also print) does not charge. You'll need to determine how many announcements you have per month and if it suits your budget to always pay for the exposure.

▸ When asked to include a headshot, make sure it's a professional image. **Do not send over a fuzzy picture you cropped from the employee's Facebook profile.** Ideally, you want to have professional headshots on file for every member of your team. When you hire someone new, bring the photographer back around and take the headshot. That's the order of operations and a great way to stay out in front of your PR.

▶ Do not wait too long with the announcement of a new hire or recent promotion. The news only stays fresh for so long. Also, a newspaper or business publication may receive a lot of announcements from the business community in a given week. Even though you submit your information on, for instance, July 10, it may not appear in the publication (or online) until the first of August. Or longer.

▶ Make no mistake: people read the business announcements section of a newspaper or business publication. In a weird way, it's the gossip section of the business community. "Oh, look who started to work at Acme law firm. I guess he quit the other firm?" Or, "Wow, Acme startup added three new people this week. It must really be growing."

○ Trust me, people will see your announcement. When you hire new people or promote your talent, tell the business community. It's positive PR and a subtle way to say, "Watch out. We're on the move."

*The publication might put your announcement in its own words, but here's what you can send over as a starting point. You'll see all three templates follow the same *fake* company as it hires and promotes from within.*

Template for a business announcement — new hire who's also a recent grad

[Name of your company] has hired [first and last name] as a/an [job title and company division if necessary; for instance, "junior engineer in the company's commercial construction division"]. [Last name] is a recent graduate from [name of college/university] where he/she received a degree in [name of degree].

[Last name] will focus on [what is the person's main role at the company? How will he/she add value?]. [Then, you can mention if the person is involved with any community groups; for instance, "He is a member of the Acme Civic League and the Acme Young Professionals Group."]

Template for a business announcement — new hire with work experience

[Name of your company] has hired [first and last name] as a/an [job title and company division if necessary; for instance, "senior engineer in the company's residential construction division"]. [Last name] spent four years as a junior engineer at [name of company] where he/she worked primarily in/with [what did the person specialize in?].

At [name of your company], [last name] will [the primary function of the new hire for your team]. [Last name] holds a degree in [name of degree] from [name of college/university]. [Then, you can mention if the person is involved with any leadership/civic groups; for instance, "He is a member of the Acme Engineering Society and Acme Nonprofit, which raises money for cystic fibrosis."]

Template for a business announcement — promotion within your team

[Name of your company] has promoted [first and last name] to the role of [job title and company division if necessary; for instance, "senior vice president. He will oversee our commercial construction division."] [Last name] spent eight years as a senior engineer at [name of your company] where he/she worked on projects like [here, you can provide names of a few notable projects the reader might recognize or find interesting, but keep the line to one sentence; for instance, "Acme Town Center, Acme Arena and the renovation of Acme Warehouse into a brewery"].

At [name of your company], [last name] will now guide the [name of team or division; for instance, "commercial construction team"] as it [here's a chance to give the reader a glimpse at what you're up to. The publication might remove the line as being too self promotional. I say include the info anyway in case the editor leaves it in; for instance, "begins several new projects in the coming year including Acme Golf Complex and Acme Fun Park"].

[Last name] holds a degree in [name of degree] from [name of college/university]. [Then, you can mention if the person is involved with any leadership/civic groups; for instance, "He is the president of the Acme Engineering Society and vice president of Acme Nonprofit, which supports childhood literacy in schools"].

54

Deeper Insight

Each announcements is short and sweet. Do not send over a person's 500-word bio. The publication does not have room for it. Plus, then you make an editor chop the 500 words down into a manageable 100–200 words. Do the editing yourself and keep the announcement brief.

Also, in case you didn't catch my drift the first time around, SEND OVER A PROFESSIONAL HEADSHOT.

That is all. Moving on.

An op-ed or guest column

You should also keep in mind the idea of an op-ed or guest column. You can generally offer the column to your local newspaper, business publication or business-focused online news outlet.

Every publication has its own rules for submission. Follow them closely — don't lose out because you skipped a step or submitted incorrectly.

A column is a great way to demonstrate authority on a topic and expose a wider audience to your business and brand.

For the template, go to the "Blog Templates" section on page 168. A well-written opinion blog post (for your own website) can double as a news column.

Two birds. One stone.

How to reach out to influencers

If you want an influencer in your space to promote or represent your product/service, the email pitch needs to be clear and contain metrics so the person takes you seriously.

Subject line: Interest in having you promote [name of company/organization]

Hi [person's first name],

I'm [your first and last name], a/an [job title] at [name of company/organization]. It's nice to meet you.

[Then, your main point; for instance, "I'm writing to see if you're interested in promoting our product, <u>Acme Sunglasses</u>. It's a line of sunglasses that are made from recycled plastic, and the entire production process is environmentally friendly."]

[Then, a section on why the influencer is a good fit for your product/service based on a specific example from his/her career — and include a link if possible; for instance, "I watched you <u>host the TV special</u> on preserving our oceans. I can tell the environment is an issue you take seriously, and that's why I think you'll appreciate what we're doing at Acme Sunglasses."]

[Next, provide your metrics/media coverage so the person can decide if an endorsement is worth it; for instance, "In the last 12 months, we have sold 25,000 pairs of our sunglasses through our online store. We also landed mentions on some of the biggest tech and shopping websites, including Acme Shop, Acme Tech and Acme Online."]

Again, we think you would be a perfect ambassador for our brand. Please let me know if you would like to talk further.

Thanks so much,

– Your first name

Email signature

Deeper Insight

The email is short but contains the necessary pieces to make a smart "pitch." You explain what you want, show you've researched the person in action, linked to what he/she has done and then round out the message with metrics/results so the person knows you're legit.

The email is designed to pique the person's interest and start a dialogue. Hopefully, the template helps you do just that.

Interview opportunities

HOW TO ASK IF YOU CAN BE INTERVIEWED ON A PODCAST

A podcast is an excellent platform to spread your message and gain followers for your business. Many podcast hosts are inundated with requests to be on their shows. How do you break through and make the host want to speak with you?

When I published *Wait, How Do I Write This Email?*, I used this template to pitch several podcasts. At one point, I sent the email to 10 podcast hosts, heard back from nine of them and appeared on seven shows. I think a 70 percent success rate is pretty darn good.

In the fictional template below, the person pitches to be a guest on *Acme Parenting Podcast*, a show to help parents raise smart, confident children.

Subject line: Interview guest: [The topic at hand; for instance, "Removing the stress from parenthood"]

Hi [person's first name],

I'm [first and last name], a/an [your job title/company and city/state]. I'm writing to introduce myself as a possible guest on [name of podcast].

>   Note: Explain at the top of the email what you want. Don't make the person hunt around for it.

[Next — and this part is crucial — tell the person about a recent podcast episode you listened to and WHY you find it notable; for instance, "I enjoy your show, 'Acme Podcast,' because you provide tips on parenting from a single mother's perspective. I thought episode 112 on extracurricular activities and managing schedules was spot on. It's so true — kids needs to learn how to budget their time even in middle school. Well said!"]

*Note: Let's unpack the section above because it's critical you understand what I did. First, I call out the podcast by name (proved I didn't send the same email to 100 podcast hosts). Then, I reference a specific episode (using the episode number), give a detail from the episode that stuck with me AND explain why I like it. In total, it's a strong intro that shows I **actually** listened to the podcast and internalized what the host had to say.*

[Then, explain why you believe you're a worthy guest; for instance, "I recently published a book I believe is a smart resource for single parents. It's called *Acme Parenting Book,* a reference guide on how to handle the most stressful situations parents face as they raise their children."]

On your show, I can talk about a variety of topics. A few examples:

- How to stay calm during a temper tantrum
- Topic 2
- Topic 3

Note: Give the host three show topics. Otherwise, the person would need to write you back and ask for topic ideas. Be one step ahead!

Here are two examples of me on previous podcasts:

- Name of podcast, half-line description of what it's about and the time in the conversation the person should jump to so he/she can hear you at your best; for instance, "Family Podcast, which discusses common family issues and how to resolve them (skip to 8:54 in the recording)"
- Second podcast info that looks the same as the one above

I hope to hear from you, [person's first name]. Thanks so much.

– Your first name

Email signature

Deeper Insight

The email above is customized for the recipient from start to finish. There's no way the person would dismiss it and say, "Oh, well he clearly sent this same email to 100 podcast hosts." The authenticity will hopefully lead to a response...which then leads to an appearance on the podcast. Mission accomplished!

How to thank the reporter/blogger/ podcaster

After you appear on the podcast, you need to send thank-you notes at two different times:

1. Right after the interview
2. Once the podcast appears online

Each time, the thank-you message will help you build on the relationship with the podcast host. You don't want to appear on the show as a one-time opportunity; the goal is to become a trusted expert the podcast host could rely on for future episodes and other special projects.

Thank-you note right after the interview

Within 24 hours, send the following email:

Subject line: Thanks again for bringing me on [name of podcast]

Hi [person's first name],

Thanks again for having me on [name of podcast]. I think you did a great job with the interview, especially [give one example from the conversation that stuck out to you; "how well you researched my blog before we spoke. You really understand my philosophy on eating and exercise, and that means a lot"].

Please keep me posted on when the podcast episode goes live, and I'll share it with my audience.

Have a great day,

– Your first name

Email signature

Deeper Insight

It's important to give an example of something the podcast host did/said that stood out to you. That's more meaningful than if you wrote, "You did a great job."

Thank-you note once podcast appears online

Hopefully, the host sends you a link to the episode once it goes live. Within 24 hours of that correspondence, reply back with the following:

[If you need to start your own email chain, the subject line can be: Thanks again for having me on the podcast]

Hi [person's first name],

Thanks again so much for letting me appear on [name of podcast]. I listened to the conversation and think it turned out great.

[Then, reference one aspect of the interview that stood out; for instance, "Hearing it again, it's clear you studied up on my business beforehand, especially my recipes for healthy desserts. I appreciate that kind of research, and it made a difference in our conversation."]

I have already shared the podcast episode on [Facebook/Twitter/my e-newsletter/wherever else you reach your audience]. Please let me know if I can do anything else to spread the word.

Thanks, and let's stay in touch!

– Your first name

Deeper Insight

Make sure to reference a moment from the interview that stands out to you. Also, note how I write "I have already shared the podcast episode." That means more than if you write, "I will share" or "plan to share." Prove you're someone who gets stuff done rather than makes promises. The podcast host will appreciate it.

Remember, you want the interview to lead to more opportunities. Use the thank-you note to build on the relationship and become a trusted resource.

How to ask if you can guest post

Let's talk about guest posts, an excellent strategy to grow a website or brand. Guest posts on sites big and small can send traffic back to your site and introduce your writing to new audiences.

I have done my fair share of "guest post" pitches, and in my experience, there are two critical pieces of the email:

1. Compliment the blog/website where you want to guest post
2. Link the reader to two or three examples of your work

Combined, the two pieces make for a complete pitch: you "give the love" and show you respect the person's work and then do a solid job sharing your own efforts.

So...the template:

Subject line: [Type of industry; for instance, "Personal finance"] expert, guest post consideration

Hi [person who runs the site],

My name is [first and last name], and I write the blog [one line about your website and why it matters; for instance, "Acme Personal Finance, which provides smart, simple personal finance tips."] I hope you're doing well.

I'm writing to ask you about guest posting opportunities.

I am a big fan of [site where you want to guest post; for instance, "ABC Money Tips"] and read your content all the time. I especially like [reference two recent posts you find worthwhile and link to them; for instance, "your articles on how to plan for retirement while a college student and 9 ways to save while working multiple part-time jobs."]

[Then, explain WHY you respect one of those links and BE SPECIFIC; for instance, "I have several part-time jobs right now and I will put tip #3 from your nine-point list to work right away — the one about how to invest $50 a month in mutual funds through a direct bank deposit. Good idea!"]

> *Note: Do you see the level of detail here? I didn't write "I think your blog posts are great." I told the person the exact piece of advice I liked ("tip #3") and how I would put it to use. The more time I spend discussing the person's work, the more he/she will be interested in what I offer.*

I wrote a blog post recently called [blog post headline that's linked to the post], and in a nutshell it's about [quick line on what the post is about; for instance, "smart tips for improving your credit score."] I would be happy to send it over as a guest post if you'd like. Here are a couple other recent posts I've done:

- [blog post headline that's linked to the post]
- [blog post headline that's linked to the post]

If you have other ideas, I am open to writing something else for [the site where you want to guest post].

Thanks, and I hope to hear from you.

– Your first and last name

Email signature with blog/contact information

Deeper Insight

First, you praise the other person and mention how much you enjoy his/her website by linking to recent content. That's a nice ego boost for the site manager and goes a long way.

Then, you link to your own content and give the person a few options. You also mention you're happy to write something else entirely if that's what the person requires. What you're "saying" is…*I will play by your rules so tell me what you want. I am the guest poster and don't call the shots.*

Summary: Make the reader feel special and be clear with your offer. That way, you stand a better chance of a positive response.

How to thank the person for the guest post

After your guest post appears online, you need to send a thank-you email to the blogger/web editor the same day.

Subject line: Thanks for allowing me to guest post

Hi [first name of blogger/web editor],

Thanks again for running my blog post about [the theme of the post; for instance, "tips to help people improve their credit scores"]. I appreciate the chance to appear on [name of website], and I have already shared the post on [wherever you have an audience and generally share content; for instance, "Facebook and LinkedIn"].

> Note: Share the post before you send the thank-you email. It's a stronger play than to write, "I plan to share the post later." Don't make promises; make things happen.

I would like to continue to be a resource for you. Can I submit other blog posts for consideration?

Please let me know.

Thanks,

– Your first name

Deeper Insight

The thank-you email allows you to build on the relationship with the blogger/web editor. You want the person, over time, to rely on you for relevant, worthwhile content. The first guest post is a way for the blogger/web editor to test the waters and see if you're legit — both the quality of your content and if you're dependable.

Hopefully, the person responds and agrees to let you submit another guest post. And then you've opened the door to a new relationship and continued exposure for your business and brand.

How to contact product reviewers

Bloggers and other influencers in your space can provide excellent exposure if they do a review on your product/service.

While it can be tough to break through and capture their attention, the best approach is an email that's focused, includes relevant information and also shows how much you value their work.

Subject line: [What's the best way to make someone stop cold and open the email? For instance, "New dieting app will change the game for good"]

Hi [reviewer's first name],

I'm [your first and last name], the [job title] at [name of company]. I hope you're doing well.

I'm writing to introduce you to [name of product/service and a one-line description; for instance, "Acme Diet App, which tracks your online behavior and suggests a meal and exercise plan that aligns with your personality"].

NOTE: Always make sure to put your main point at the beginning of the message so the person knows what you want right away. And include a link to the product/service.

I think you'll appreciate the app because [why would the person find it relevant?; for instance, "of the great reviews you provide for Acme Tech Review"]. [Then, reference a recent review, link to it and explain why you liked it — and be specific! For instance, "I especially like how you reviewed the Acme Smart Scale and checked your own weight every two hours throughout the day. That was a clever way to see how accurate it is."]

Here's a bit more info on Acme Diet App:

- Three bullet points that provide further explanation on the product/service. Lean on metrics whenever possible; for instance, "Since we launched the app three months ago, we already have 35,000 downloads."

- Or the metric can refer to success rates; for instance, "In a recent survey, our users found the app developed a proper meal/exercise plan 77% of the time."

- Third bullet with more information like product features or other metrics.

Please let me know if I can provide additional information on [name of product/service]. Again, we value your work and would appreciate a review.

Thanks,

– Your first name

Email signature

Deeper Insight

Step back and look at the big picture here. You approach a product reviewer because you know the person has clout and influence. That's why a great way to nab the person's attention is to compliment his/her previous work.

Combined with concrete details on your own product, you make the strongest pitch possible in a small amount of space.

If the reviewer doesn't answer after 48 hours, you can reply to the email and write:

Hi [reviewer's first name],

I want to make sure you saw my original email that requested a review of [name of your product/service]. Please let me know and thanks again.

- Your first name

If again you receive no answer, consider calling the person by phone and have your "pitch" ready.

Prepping for the interview

WHAT TO WEAR

There are three styles of dress appropriate for an on-camera interview, and none of them is "casual." You need to step it up for an interview and look presentable. Here are your options.

1. Company uniform
2. Business casual
3. Dressy

Even if your company doesn't require employees to be "dressy" (ex: suit and tie), you still need to wear business-appropriate clothes.

If you have questions about style of dress, talk with your team and/or management. Do NOT gamble on what to wear or look sloppy. How you dress speaks volumes and, if your clothes are distracting, they will overshadow what you have to say.

Oh, and remember to brush your hair and pick the broccoli from lunch out of your teeth.

How to plan out your talking points

Whether your interview is live (ex: in the TV studio) or recorded (ex: reporter interviews you in your office), you need to have talking points worked out in your head.

If you stumble a bit in a recorded interview, you can always ask the journalist if you can start the sentence over. Obviously, in a live setting you don't have the same luxury.

Either way, plan out what you want to say.

On your own or in conversation with your team, discuss the following:

▸ What main points do you want to highlight?

▸ If you only had ten seconds to convey your central point or argument, what would it be?

▸ Is there information you're not willing to share? Topics that are off limits for privacy concerns or other business reasons?

As well, consider practicing a few key lines so you're comfortable saying them "in real time."

Finally, someone else on your team could act as the journalist and conduct a mock interview so you're in the right frame of mind.

Why you need to think about the scenery

If a journalist plans to visit your office or another location related to your business, think about the images and visuals.

Ask yourself:

▸ Will there be quality video for the journalist to capture? For example, visuals of people doing work at your company — whatever that work may be.

▸ Do you need to prep anything so the visuals are the best they can be?

▸ Is everything clean and organized?

Then, for the interview itself, make sure the backdrop doesn't distract from the conversation. At the same time, a plain white wall is boring.

The journalist may create a backdrop for you, but you should think ahead of time about spaces that are well lit and quiet.

Final point: Give 'em a break

The media often receives a bad rap in, well, the media. People complain about news coverage (left, right or center). And business owners may gripe about the stories the media chooses to report.

"Hey, why did Channel 5 interview Acme Corporation? We do the same type of work and Tech Corporation, and we're better!"

As someone who spent time in newsrooms as a reporter and then consultant, I can tell you it's no easy task delivering the news every day. Journalists — print, TV and online — work long hours and increasingly have to file multiple stories in a single day. They move fast, do their best and can't come through for everyone every time.

If you don't land news coverage, that's OK. You will have more opportunities. The key is to build relationships with the press so they come to you when they need a quote or story subject.

But in general, cut the media a little slack. Journalism is perhaps the most thankless line of work, but we'd all be lost without it.

My Journey: Part 2
That time I checked my website traffic and it led to a book idea

I wasn't looking to write a book. The thought never crossed my mind, really.

In the early days with my blog, my idea was to create a news column that offered career advice based on the actions of people in the headlines. I called it News To Live By.

Then, about 18 months into my blogging journey, I noticed an interesting trend in my site traffic (thank you, Google Analytics).

And I knew, from that moment on, I needed to be an author.

What happened?

I wrote a blog post in November 2013 called "Five Ready-to-Use Templates for Tricky Job Situations." I thought a series of email guides was a cool idea and useful reference.

I published the post, put the content on Facebook, Twitter yadda yadda and moved on to the next piece.

Six weeks later I poked around Google Analytics, and I noticed something interesting.

The "Templates" blog post had about 300 views a day — every day. Most of my other posts (career advice based on the news) had little web traffic so I didn't understand the spike in the "Template" one.

Then it dawned on me. People had been searching the web for help with job-related templates and started to find my post.

I thought, "Wait a second. Maybe writing step-by-step guides has traction."

From that day forward, I decided to pivot and devote my blogging to practical writing guides. I wrote over 75 posts on resumes, cover letters, LinkedIn, networking and job interviews.

Over another 18-month period, I had built up a solid base of content and the web traffic grew along with it (over one million views to my content). All organic traffic, no ads.

That's when I knew I had a book idea on my hands.

The book, *Wait, How Do I Write This Email?*, helped to establish me as a go-to expert on business communication skills. I made a name for myself within several career service-related organizations (see page 32 about joining associations) and began to develop an audience around my niche topic (see page 178 about e-newsletter copy strategies).

I believe my "big" idea (book on writing guides) materialized because I was (and still am) 100% committed to my craft. I pursued my 10,000 hours and in turn was shown my "a-ha" moment.

If you want to "change the game" in your industry, work on your own skills — in obscurity — for **at least** three years (and maybe more).

When you step back into the light with your innovative idea, the world will be ready. It always is.

In the next part of "My Journey," the book begins to take shape.

Chapter 3
Networking and Outreach Emails

Relationship building

HOW TO MEET WITH SOMEONE TO ASK FOR ADVICE

Important people in business are impossible to reach...until you ask for their advice. Then, many are willing to stop in their tracks and spend time answering your questions. Remarkable how that happens. :)

Here's a template to ask for a meeting and seek someone's wisdom (and ultimately, add him/her to your network).

Subject line: [Your industry] professional who could use your advice

Hi _____,

NOTE: Refer to my chart on page 16 to determine the best way to address the person.

I hope you're doing well.

I work for/own [name of company], a/an [one sentence about your business; for instance, "a new children's dental practice in Columbus"].

> NOTE: Link the name of your company to the website.

[Then, explain the purpose of your email and why you would value the person's insights; for instance; "I'm writing to seek your advice as I grow my small business. As someone who has spent many years in the Columbus business community — albeit in a different medical specialty — I think you could provide insights that would make a big difference for me."]

[If you write someone you don't know, a smart way to make the pitch even stronger is to google his/her name and read something the person has written or been quoted in; for instance, "I read an *Acme Daily News* article in which you were quoted about the importance of entrepreneurship. Your words stuck with me, especially the part about how you need to push through when the going gets tough."]

> NOTE: It's not enough to tell the person you liked the article or interview quotes. You need to explain WHY you enjoyed the article and reference a specific part that stood out to you.

Please let me know if you have time for coffee over the next several days.

Thanks, and I hope to hear from you.

– Your first name

Email signature

Deeper Insight

The two big keys with the "advice" email.

1. Tell the person at the start of the email what you want. That way, you're direct and to the point.
2. Explain why you seek the person's knowledge and, when possible, be specific about how much you respect what he/she has to say.

How to thank someone after the networking meeting

I recommend you send a thank-you message within 24 hours to recap the meeting and build on the relationship.

Subject line: Thank you again for meeting with me

Hi [Mr./Ms. or first name depending on how you wrote the message to ask for a meeting],

> *NOTE: Refer to my chart on page 16 to determine the best way to address the person.*

Thanks for meeting and giving me advice about [topic at hand; for instance, "how to grow my dental practice."] I value your time and all you had to say.

[Then, give a specific example of the advice the person told you; for instance, "I appreciate how you explained the best way to make a cold call. I will put that advice to use this week as I have several people I need to reach."]

Of course, I am happy to help you any way I can. Don't hesitate to ask.

Thanks again, and I hope to see you soon.

– Your first name

Email signature

Deeper Insight

Repeat the person's advice back to him/her — it's an ego boost and will prove you listened. Plus, the strategy helps you build on the relationship.

Also, if you and the other person have plans to see each other again, reference your upcoming meeting rather than "I hope to see you soon."

How to keep in touch even if you don't do business together

There are people in your network who need to hear from you every now and then. They may be influencers or connectors who can open the door to different opportunities. You always want the door cracked open rather than locked shut.

It's tough to gauge how often to write these "connectors," but the best rule of thumb is to reach out when you have big news or an announcement of some kind.

In the example below, the email recipient is a fictional investor who's always looking to fund the next big idea. The sender has known the person for a couple of years and feels now is a smart time to check back in.

Subject line: Checking back in + big news to share about [name of your company]

Hi [person's first name],

I hope all is well. [Then, ask the person a question about his/her career; for instance, "How is everything going at Acme Research Labs? Last time we spoke, I know you were preparing to fund a new ridesharing platform. I hope that project is still on track."]

I'm writing to check back in and let you know [your big announcement; for instance, "my company, Acme Tech, which developed a program to allow cars to detect bikers nearby, was voted top startup at the recent Omaha Startup Challenge. We beat out 27 other companies and won the $5,000 grand prize"].

[If you want to provide additional information, do so here but keep it brief and include links, when possible; for instance, "Here's more information on how the technology works"].

It would be great to catch up further if you have a few minutes by phone. Please let me know when you're available.

Thanks, and I'll talk to you soon,

– Your first name

Email signature

Deeper Insight

Begin with small talk and then explain the "big" news in a clear and concise manner. And then ask for a phone call to push the discussion further.

If you feel, given the person or situation, that a phone call is too pushy, then you can write, "I want to make sure I passed along the good news. Thanks, and I hope to hear from you."

How to ask a co-worker for networking time

If you would like to meet with a co-worker to pick the person's brain, network or learn new skills, you need to send an email that's personable yet to the point.

The email is particularly useful for interns or younger employees. Networking with your team is a great way to build rapport and become smarter at the work you do.

Subject line: Hope to learn more about [topic at hand; for instance, "Acme data servers"]

Hi [person's first name],

I hope you're having a good week so far.

When you have time, it would be great if you can [what you want from the person and why; for instance, "teach me more about our new Acme data servers. I know you understand the technology, and I need to bring myself up to speed."]

75

[Then, set time parameters but keep them flexible; for instance, "Let me know if you have a few minutes to spare over the next several days."]

Thanks in advance,

- Your first name

Email signature

Deeper Insight

Don't push for a meeting right away or leave it open-ended with "whenever you have time." By asking for "a few minutes over the next several days," you show you want the meeting soon but not immediately.

How to announce you started a new business

When you launch a new venture, you'll want to tell everyone, right? Before you spread the news, make sure the email contains the right info.

If the email goes to one person (rather than a group), here's the template.

Subject line: Introducing my new [company/organization], [specific name of company/organization]

Hi [person's first name],

I hope you're doing great. [Then, one line of small talk about the other person before you discuss your new business; for instance, "How is everything going at Acme Real Estate? Have you been busy with new leads this summer?"]

[Then, explain right away the purpose of your message and link to your new website; for instance, "I'm writing to tell you I started a new company, Acme Arts Discovery, a program for special-needs children in Portland that provides hands-on art instruction."]

[Then a line to explain why you felt the need to start this new venture; for instance, "I had such positive feedback from the arts classes I led last summer that I decided to go a step further and make an entire business out of the idea."]

[Then, what do you want from the person? You should always have a call to action and a way to engage; for instance, **"Here's the upcoming class schedule. Do you know of any families who might want to participate?"**]

> Note: You should consider highlighting or putting in bold the "call to action" so the person doesn't miss it.

[Finally, allow the person to provide other ways to promote your business and help you network; for instance, "If you have other ideas on how I can spread the word, please let me know."]

Thanks, and I hope to hear from you.

– Your first name

Your new email signature

Deeper Insight

Make sure you have everything ready with the new venture before the email goes out. That means your new website must be live with all the kinks worked out. And don't forget to give the person the floor to brainstorm ways he/she can help. The person might come up with an idea or connect you to someone you never even considered.

If you'd prefer to send this email to a group of people, here are the modifications:

- Send the email to yourself and BCC everyone else.
- Open the email with "Hi everyone,"

- Don't include the lines of small talk. Instead, write "I hope you're doing great."
- The rest of the email flows the same whether it's to one person or a group.

How to introduce yourself to a company for the first time

The "cold call" email is one of the most challenging outreach messages for a business owner. The recipient doesn't know you or trust you. How do you win the person over?

The key is to make the email personal and keep it brief. You want to appear authentic, open, honest and as someone who knows how to "get to the point."

Subject line: Smart new resource for [name of person's company]

NOTE: There are many different approaches for the subject line. You may have seen slick ones like "Looking for the right contact person," "Trying you again" or "Following up." These phrases give the impression you have met the sender before. It's an empty gesture and once we realize it's a canned message with no authenticity, we're turned off. That's why I always want to be transparent and honest.

Hi _____,

I'm [first and last name] with [name of company], [and then a short line about what the company does so the reader has context; for instance, "a maker of organic treats for dogs and cats"].

I hope you're doing well.

I'm writing you because I think [name of person's company; for instance, "the team at Acme Pet Supply"] would like to learn more about [name of company; for instance, "Acme Pet Treats"]. [Why should the person care? Why does your business matter? For instance, "The product is healthy, safe and selling out routinely on our website"].

NOTE: In the above section, you need to explain the purpose of the email and why your company matters. People are busy so make your main point high up in the message to prove your worth.

[Then, one sentence to show you studied the company's website. It's a powerful way to prove you didn't send the same email to 100 business owners. Be as specific as possible. For instance, "Also, I want to tell you I enjoyed the photos from the recent Bark Bark 5K race you sponsored. The chihuahua with the running shoes? Too funny!"]

NOTE: Remember, it's not only about selling products. It's about building a relationship.

[Now, show people results. Here, you need to provide an example of a recent success with data to back you up; for instance, "We began to sell Acme Pet Treats online in January 2017. In four months, we sold 4,500 units with limited social media marketing. Most of the buzz has been through word of mouth."]

NOTE: However you define and quantify success in your business, provide one clear example. Make sure the case study has metrics. Otherwise, you have no proof.

A bit more information on [name of company]:

NOTE: Ask yourself, "What else does the person need to know to trust our business?"

- All of our treats are made with organic products.

- We can fulfill any order up to 1,000 units within seven days.

- Like Acme Pet Supply, we are also located in the Phoenix area and can provide in-store samples.

Please let me know your thoughts. I'm happy to answer any questions and hope we can start a conversation.

Thanks,

– Your first name

Email signature

Deeper Insight

Many of us are accustomed to deleting a "cold call" email before we even read the first line. That's why it's critical to make the message customized and personable.

Notice how I put the company's name in the subject line and also reference something from the company's website that stands out. The approach makes it tough to say, "This person is a sleazy salesman" because, well, there's nothing sleazy about it.

It's tough to win over a stranger in a "cold call" email, but you give yourself the best odds with a message that's designed to gain someone's trust as much as sell a product/service. **The key is to mention a specific project from their website, link to it and explain why you respect it.**

How to introduce a client to another business associate

Let's say you want to connect a client to another person in the business community.

In that case, you should send an email that introduces everyone and does so in a way that provides relevant information for each party.

Subject line: Making introductions for [name of your company] client

Hi _____,

Good morning/afternoon.

I hope all is well at [name of business associate's company/organization].

I'm writing to introduce you to [first and last name of your client; for instance, "Theresa Brown"]. He/she is a/an [job title] at [name of company], and I thought I would connect you both because [what's the purpose? For instance, "Theresa's company makes custom signs and banners. Perhaps she could help with your grand opening celebration?"].

[Then, introduce your client to the business associate; for instance, "Theresa, my friend John owns a chain of car washes in Detroit called Acme Car Wash. He has six locations in the region and plans to open the seventh in April."]

I hope you two can connect and find ways to work together.

Thanks,

– Your first name

Email signature

Deeper Insight

It's important to provide a short intro about both people and link to each company website. That way, the people can check each other out before they reply to the email chain you started.

The best networkers often send these kinds of emails. They think beyond themselves and want to help others succeed. Plus, they strengthen their own network by bringing their various connections closer together.

A win-win situation all around.

How to ask a current client to make an introduction for you

Trusted clients can often help you win over new business prospects. They can vouch for your service and also connect you to new people without the need for a "cold call" email.

Subject line: Hoping you can connect me with [name of person at name of company]

Hi _____,

Good morning/afternoon.

I hope all is well at [name of company/organization]. [Then one line for a little small talk; for instance, "I see the big Acme Charity Gala is coming up in November. I'm sure you have your team ready for the challenge once again."]

I have a small favor. Are you able to send an email and connect me to [name of person; for instance, "Derrick Hilliard who runs Acme AeroTech?"] [Then, explain why you want the connection; for instance, "I know you have worked with Derrick before, and I would like to discuss partnership ideas. He's a busy guy so maybe an introduction from you would move things along."]

Please let me know if you can help. I can send you brief information on [name of your company] for your intro email.

Thanks,

– Your first name

Email signature

Deeper Insight

The key is to ask for the assistance but not provide details on your company *until the person agrees.* Once the person says, "Sure, no problem," then you can send over information about your business.

Your follow-up can look like this:

Great. Thanks a lot for the help.

Feel free to use the information below to write your email. And please cc me on the message you send Derrick.

———

Derrick, I want you to meet Brent Parsons who handles the bookkeeping for our team at Acme Restaurant Group. Since 2009, Brent has owned Acme Professional Bookkeeping, which works with dozens of small to mid-size businesses in the Tampa Bay area.

Brent is dependable and works fast. Here's a list of his services.

I hope you will allow Brent to explain his offering further.

Deeper Insight

The person may put the "introduction" email in his/her own words, but you have provided a framework, which is helpful too. Also make sure to include one to two links so you control what the email recipient reads about you.

Client outreach

HOW TO OFFER A NEW PRODUCT/SERVICE TO EXISTING CLIENTS

When you want to offer a product/service to existing clients, the best strategy is to tell a story of how the product/service already made a difference for someone else's business.

As you write, ask yourself: how did my product/service solve another company's problem? And how can I turn that experience into a short story?

I used this approach when I helped a large IT firm in Virginia pitch a new product to its clientele. The email went to 200 clients and lead to new business with 20 of them. One email that upsells current clients **10 percent of the time?** Yes, please!

Subject line: [A line that makes the reader think, "Hmm, am I falling behind?" For instance, "Is your business prepared for the next big virus?"]

Hi [person's first name],

I hope you're doing well.

Let me tell you a quick story about [name of client; for instance, "Acme Grocery"], [and then a short line about what the company does so the reader has context; for instance, "the largest grocer in the Southeast"].

[Then, explain what happened in one to two sentences; for instance, "In March 2016, a dangerous internet virus called Mega Virus ran loose and threatened every piece of Acme Grocery's internal data".]

Note: Use the opening section to introduce the story and grab the reader's attention.

[Then, explain the consequences if the problem wasn't solved; for instance, "Once the virus infects your system, you either pay a ransom to the hacker or reset your computer to factory settings and lose everything. Clearly, both are terrible options."]

[Next, tell the reader how your product/service stepped in to save the day; for instance, "Fortunately, Acme Grocery uses Acme IT for security monitoring services. As soon as our team learned of the virus, we added security rules and firewalls to the grocer's network to prevent the virus from reaching its computers. Today, Acme Grocery goes about its business and never has to worry about Mega Virus or losing critical files."]

[To wrap up the story, include a quick-hit line; for instance, **"Crisis averted."**]

*Note: Put the quick-hit line **in bold.***

[Then, ask the reader the "Are you prepared" question in italics; for instance, "Is your business protected from the next big virus? Are you backing up your data every single day?"]

[Next, share your product/service offering(s) but keep it short; for instance, "Please read the brief descriptions of our security services to learn more.

1. **Proactive Monitoring:** We keep watch 24/7 over your IT components so server, network and desktop outages don't happen. Period.

2. **Network Patching:** Your IT components receive the latest security patches from each hardware/software vendor. Like Acme Grocery and Mega Virus, we make sure the bad guys never even have the chance to sneak in.

3. **Security Monitoring:** We also keep close watch on every security patch so they're up to date and not vulnerable to attack. It's not enough to install a patch and walk away; it must be monitored day after day. No exceptions."]

[Finally, explain once more why your product is necessary; for instance, "Hackers and viruses are a real and constant threat to businesses of every size. Your computer systems must be protected and ready to defend themselves from the next attack."]

[And to conclude, urge the reader to contact you; for instance, "Again, I'm happy to talk about how Acme IT can help. Please email or call me at your convenience."]

Thanks,

– Your first and last name

Email signature

Deeper Insight

Don't become overwhelmed by the template here. I know it looks like a lot of writing, but that's because I included several NOTES and layers of explanation.

Remember: how would anyone — even a current client — trust that your new product/service works unless they see how it's used in real time? **The short story does the selling.** And after the story, provide product/service details (but don't go on too long, one or two sentences for each aspect) and wrap up the message.

How to follow up after you send over a proposal or contract

You're on the verge of a business deal, but the proposal/contract is stuck in email limbo. Did the message go through? Follow up and find out.

Subject line: Following up on [name of project] [proposal/contract/agreement]

You: Hi, my name is _____, and [earlier this week/last week] I sent over a [proposal/contract/something else you need reviewed or signed]. I am writing to make sure you received the document.

As you have time, please let me know.

Thanks again,

– Your first name

Email signature

Deeper Insight

If you need a client to sign a time-sensitive document, don't sit there and wait for an email — a phone call is the most expedient way to receive a confirmation.

How to request testimonials and when

After someone experiences your product/services, it's important to ask for a testimonial. The challenge is to determine the best time to request one.

In some cases, you might ask a client/customer right away through an automated email. But if you want to make the request personally, it's best to ask for the review within two to three days of when you completed the work.

It can sometimes feel too pushy if you finish the project and then immediately ask for a testimonial. On the flipside, if you wait several weeks, the person's memory won't be as fresh. Wait a couple of days and then send the following email.

Subject line: Request for a short testimonial

NOTE: Put what you want in the subject line. Don't bury it with a line like "Quick question." Be up front about it.

Hi _____,

Good morning/afternoon.

If you have a chance today, can you please provide a short review on [Where will it go? Your website? Another review site? For instance, "your bathroom remodel for our Testimonials page?"]

It would mean a lot to have a review from you. [And then one line specific to this client so the email feels special and authentic; for instance, "We're proud of the remodel we completed on your master bathroom — especially the custom work on the two sinks!"]

Your testimonial doesn't need to be long — a few sentences will be fine.

Here's the link to leave the review.

Please let me know.

Thanks,

– Your first name

Email signature

Deeper Insight

You have to be delicate with the request here and not demand a review. Notice how I wrote phrases like "If you have a chance…", "short review" and "a few sentences will be fine."

At the same time, I gave the person a time parameter ("today"). Hopefully, the gentle suggestion encourages the person to do the review now and not forget about it.

How to wish someone well in the next adventure

As people start new chapters in their career, it's a smart relationship-building gesture to bid them farewell.

Subject line: Wishing you all the best

Hi _____,

Best of luck with your big move to [a new job, a new city or both; for instance, "your fellowship at the Acme Institute in Washington, DC"]. I'm excited for you and know you will make your mark in no time.

We will miss you at [name of company or organization where you spent time together]. I always appreciate [what sets the person apart? For instance, "your positive attitude and willingness to help others"]. [Then, a specific example of how the person helped you; for instance, "Plus, I'll never forget running around town with you as we tried to find the right hotel for the Acme Training Series. Third time was the charm!"]

I hope to see you soon, and let's stay in touch.

– Your first name

Email signature

Deeper Insight

If you want to maintain a business relationship, take two minutes and wish someone well. The best networkers always think about other people first.

How to check in with donors or board members

When your organization has good news to share or you need leadership to be aware of an announcement, check in via email.

Subject line: Big news to share at [name of company/organization]

Hi everyone,

Good morning/afternoon.

I'm excited to report that [What's the big news? For instance, "thanks to the huge popularity of our 10K charity run, we have surpassed our fundraising goal of $25,000! The grand total: $33,561. Here's an article from *Acme Daily News* about the event.].

> *NOTE: Put the big news right at the top. Don't make the person scan the email and look for it. And if you obtained media coverage, let your board/donors see it!*

[Then, a few details on how the successful project came to be; for instance, "We all owe a big thanks to our staff and the 50+ volunteers who made everything happen. We even had glitches with online registration a week before the run, but we still managed to include every person who wanted to participate."]

[Then, explain what the success has allowed you to do; for instance, "Due to the overwhelming success of this event and the additional funds raised, we will be able to plan another charity run in the spring and potentially increase our fundraising goal to $50,000."]

[Then, a thank you to your board/donors and a call to action; for instance, "Thank you to our entire board for having the vision to stage the race and the confidence in our organization to do the job.

Here's a post you can share on <u>Facebook</u>, <u>Instagram</u> and <u>Twitter</u>. Click on each social media link to find the post. We would appreciate your help spreading the word about our success."]

> NOTE: Whatever social media channel(s) you prefer, have ready-made social media content so the person can easily click and share.

[Finally, if there's any other additional information, add it here and place in bold if you need to make sure people see it; for instance, **"Remember, our next board meeting will be Sunday, March 7 at 6 p.m. in the Acme board room. I hope to see everyone then."**]

Thanks so much,

– Your first name

Email signature

_____

Deeper Insight

The email above helps you in two ways. For one, you pass along good news so others can help you share it. Beyond that, you keep important people up to date on the organization. Then, they feel "in the loop" and part of the process.

For more "in the loop" emails, go to page 111.

How to reach out to a potential partner organization

Business partnerships can bring new skills/services to your clients and increase the number of referrals.

To ask about partnering up, the email must look and feel authentic so the reader trusts you. Avoid the cookie-cutter, impersonal approach at all costs.

In the example below, a builder hopes to partner with a landscaping company.

Subject line: Interest in partnering with [name of your company/organization]

> *NOTE: As best you can, try to send the email to a person rather than a general "Contact" page. If you can't locate a person, then open the email with "Hi [name of company/organization] team."*

Hi [person's first name],

I'm [first and last name], a/an [job title] at [name of company/organization; for instance, "an award-winning homebuilder in the Salem area"]. I hope you're doing well.

I'm writing to introduce myself because [the reason; for instance, "there may be ways for us to work together on upcoming [type of work you do and why you hope to connect; for instance, "home construction projects. We excel at building custom homes and are in search of a new landscaping partner"].

I read up on the [name of their company/organization] website and like the work you do. In particular, I thought your recent [find something notable on the website and link to it; for instance, "landscape job in the Acme Meadows community was top notch and showed a great deal of creativity"].

> *NOTE: If you can't find an example of a recent project, then tell the person you respect the company's philosophy (About Us page) or appreciate the team's wealth of experience (Our Team page). But it's always best to discuss a recent project.*

[Then, a quick intro on your own team with a few highlights; for instance, "Our team at Acme Home Building has designed and built homes throughout

Salem since 2003. We have twice been named 'Best Builder' in Salem and our innovative designs often win industry awards.

Here's a look at our upcoming community, Acme Park West"].

Please let me know if you're interested in a further conversation. Again, we think highly of [name of their company/organization] and believe we can create great work together.

Thanks,

– Your first name

Email signature

Deeper Insight

It's critical to make the email customized to the recipient. That's why you need to reference a past project from the company/organization website. It shows you did your research and meant what you said.

Also make sure you link to at least one of your own projects to showcase your best work.

How to invite new business prospects to an event

If you plan a company event and want select clients or others in your network to attend, you need to make the email outreach unique for each person. Also, if the event is significant enough, you may consider a printed invitation followed up with a personal email.

Subject line: Inviting the [name of company] team to our [type of event; for instance, "grand opening"]

Hi [first name of person or name of company and then "team"],

I hope you're doing well!

Note: If you have a relationship with the person or team, add a little small talk in place of "I hope you're doing well."; for instance, "I hope everything is going well with your office renovation. Will everything be completed soon?"

I'm writing to invite you to our [company event; for instance, "Hill Valley location grand opening"] on [day/time]. The address is [list the address].

[Then, a line on why the grand opening would be of particular interest to the person/team; for instance, "I think your team will enjoy this grand opening because we will have several guests who work in interior design. Maybe one of them could help decorate your new offices?"]

Note: The above section is one more way to make the email custom and authentic to the recipient(s). Always think about the value you can add to someone else's business!

[Then, a few more details about the event; for instance, "We'll do the ribbon cutting at 4 p.m. if you can make it by then. After the ribbon cutting, we'll have heavy hors d'oeuvres so come hungry and ready to mingle!"]

Thanks, and please let me know by [RSVP date] if you can make it.

– Your first name

Email signature

Deeper Insight

The email strategy here will encourage the person/people to respond. Certainly, not everyone will be able to come; we all have conflicts now and then. But your outreach is warm, genuine and conversational.

As well, the email approach will land responses faster and limit how often you need to follow back with, "Checking to make sure you saw my original invite. Let me know if you can come!" Those emails are annoying to send, I know.

Make sure you take good group photos

At the event itself, you MUST take photos of the guests. In particular, make sure you snap pictures of people you invited — or as many as you can.

For one, the photos make the invited guests feel special. The pictures also become smart social media material. That's because you can create a photo gallery, send it around to invited guests and encourage people to share the photos (particularly the ones with them in it) on their social media feeds.

If the event is significant enough, it's worth the investment to hire a professional photographer. High-quality images have far more marketing potential than ones captured on your phone or a camera with a lower-quality lens.

How to thank someone for attending an event

After the event, it's a smart move to thank people who attended — either key people you hope to do business with or everyone who came (if it's not too onerous a task).

Yes, it requires extra effort, but the best thank-you notes contain a special reference to a moment the two of you shared or something specific that happened.

Subject line: Thanks again for attending [name of your event and be specific; for instance, "Hill Valley grand opening"]

Hi [first name],

Thanks again for coming to the [name of event; for instance, "Hill Valley grand opening] this past week. I'm glad you were able to attend and [the reason you're glad; for instance, "see our new store firsthand"].

[Then, the special reference; for instance; "Plus, wasn't the catered food delicious? It was so funny how we 'fought' over the mini grilled cheese sandwich. Next time, it's mine!"]

[If you don't have a moment the two of you shared, be specific about another topic; for instance, "I know you're not the biggest fan of country, but I hope you still enjoyed the twangy country singer who performed!"]

[Then, a call to action; for instance, "It would be great for our teams to work together as the Hill Valley location gets underway. Let's stay in touch."]

Thanks, and I hope to see you soon.

– Your first name

Email signature

Deeper Insight

Someone took time out of his/her day to attend your event. That person *should* receive a thank-you note. It's a big deal. On top of that, you make the email personal and authentic (ex: the reference to the mini grilled cheese sandwich). That little "wrinkle" in the message makes it memorable and helps you build on the business relationship.

If you can carve out the time, I recommend a handwritten thank-you note instead — particularly for select people who could aid your business. Again, I know it takes time and you have 1,000 other things to do to run your operation. But relationships are everything, and no one forgets a handwritten note.

How to keep clients in the loop

HOW TO RELAY THAT YOU'RE TROUBLESHOOTING A PROBLEM

(Most) clients don't expect perfection, but they do insist on being kept "in the know." So often, a short, simple email that explains the situation or provides an update is enough to satisfy a client and maintain a positive relationship.

Rule of thumb: when in doubt, over-communicate.

Hi _____,

Or "Hi everyone," if you write a group

I want to let you know we are aware of [name of specific problem; for instance, "the mistake with the shipment of hard drives to California"]. Our team is working on the problem now and will continue to provide updates as they come available.

[Then, any additional information the person may need to know. If you have several different points, consider putting them in a bulleted list; for instance:

"A few more pieces of information at this time:

- We can confirm the shipment of monitors to Ohio has arrived on time.

- Steve Maxwell, our warehouse manager, is on vacation starting tomorrow. Your new point of contact will be Tina Johnson, assistant warehouse manager. She can be reached at [email] and [phone number].

- If you have any questions after business hours, please call our 24-hour hotline at [phone number]."

Thanks, and please let me know if you have any further questions right now.

– Your first name

Email signature

Deeper Insight

Notice how I refer to the issue by name ("shipment of hard drives to California"). I don't write "we are aware of the issue." That's too vague. Make sure the email recipient knows exactly what you reference.

How to check in after too much time has passed

The next email template is ideal when it feels like the client needs to hear from you again. Otherwise, the client may feel in the dark and begin to wonder, "What am I paying this person for?" You don't want to reach that point, and an email like the one below is a way to avoid it.

Hi _____,

Or "Hi everyone," if you write a group

I want to update you on the status of the [name of project; for instance, "Acme Cafe restaurant redesign"].

Here's the latest:

> *NOTE: If you have a lot of information, use bullet points. Do not lump everything into a big paragraph.*

- We are on track for the investors to do a walk-through on Monday, June 23 from 3–4 p.m.

- The lighting for the bar has been ordered. **John,** when will it arrive?

- We are still waiting on the dimensions for the front door and have not ordered materials yet.

Thanks, and please let me know if you have any further questions right now.

– Your first name

Email signature

Two points about the bulleted list:

1. I put the most positive news at the top and the "bad" news (not yet ordered the materials) at the bottom. It's always good to lead with your best information.
2. If you want to grab someone's attention, put his/her name in highlight or bold.

Why you should consider "Recap on a Friday" emails

Friday afternoon is an ideal time to check in with the client and provide a full update on where a project stands. Then, everyone heads off for the weekend and feels confident about what's going on.

Here's the same email from the previous template ("How to check in after too much time has passed") but written around 12 or 1 p.m. on a Friday. Don't send the email at 4:45 p.m. because it's unlikely people will respond (or they may not see the message at all).

Hi _____,

Or "Hi everyone," if you write a group

Before we hit the weekend, I want to pass along some updates and catch everyone up on the [name of project; for instance, "Acme Cafe restaurant redesign"].

Here's the latest:

NOTE: If you have a lot of information, use bullet points. Do not lump everything into a big paragraph.

- We are on track for the investors to do a walk-through on Monday, June 23 from 3–4 p.m.

- The lighting for the bar has been ordered. **John,** when will it arrive?

- We are still waiting on the dimensions for the front door and have not ordered materials yet.

Please let me know if you have any questions at the moment. I will be in touch again on Monday with any further updates.

– Your first name

Email signature

Deeper Insight

Give people a chance to respond to you today (Friday) but also make clear you will be in touch on Monday too.

How to handle an issue/inquiry

When something goes wrong, it's critical to stay in contact with the client. People generally understand "accidents happen," but what makes them mad is a lack of communication.

What's going on? Is it being fixed? Be sure to answer those questions in your email correspondence.

Subject line: Update on the [issue at hand; for instance, "typo in the marketing brochure"]

Hi [first name of client],

I want to let you know we are working to fix the [issue at hand; for instance, "typo on the front of the tri-fold marketing brochure"].

[Then, further details on the situation; for instance, "I also called the printer and told her to wait on printing the 5,000 copies until we have a corrected file from the designer."]

[Then, the next steps; for instance, "As soon as I have a new file, I will let you review before it goes to the printer."]

I will keep you posted,

– Your first name

Email signature

Deeper Insight

If you feel the need to apologize, you may consider adding a line like, "We apologize for the mistake and are working hard to correct it."

Also, let's say in the above situation, it takes the designer a bit too long to correct the brochure. By the end of the day, you don't have a new file to show the client.

That's when you want to send another email (you can reply to an ongoing chain) to keep the client updated.

For instance:

Hi [first name of client],

I want to give you a quick update. The designer has not sent me the updated file, but as soon as he does, I will share it with you.

Thanks for your patience,

– Your first name

Deeper Insight

Would the client be happier if you sent over the new file? Of course. But you're doing your best to maintain open lines of communication, and the client will appreciate the extra effort.

Handwritten note opportunities

In a world of rapid-fire electronic communication, a handwritten message is a breath of fresh air.

Handwritten notes are reserved for the times when you want to say "Thanks" in a bigger way or make a lasting impression as it relates to your business.

Below are several ideas and templates to help you do the most with good ol' pen and paper.

HOLIDAY CARDS

Holiday cards are a great time to check in with your clients/vendors and extended professional network.

There are countless directions to take your holiday card in terms of creativity. All I will say is this: a clever, memorable approach will help the card stick around the recipient's office rather than go into the trash.

What will make your card "fridge worthy"? If the company saves your card, it becomes smart PR for your business.

Thank someone for an opportunity/ experience

Name,   Month/Day

Thank you again for [the opportunity/experience; for instance, "inviting me to your company's annual charity gala as a guest at your table"]. [Then, why it meant a lot to you and something specific that happened; for instance, "It was great to meet your entire division, and I had a particularly good time with Ronda Jenkins. She's also big into running, and we had plenty to talk about"]. Again, I appreciate you thinking of me, and I hope to see you soon.

                                                    – Your first name

Deeper Insight

Make sure you reference a specific moment from the event/opportunity. It will make the note more personal and allow the person to feel even better about inviting you.

The thank-you note should go in the mail within 24–48 hours of the event/opportunity.

Congratulate a client/vendor on a big moment

Name,   Month/Day

Congratulations on [the big moment; for instance, "winning Best Small Business of the Year from the Omaha Chamber of Commerce!"]. I know your team worked hard and you should be proud of the accomplishment. [Then, something additional about the big moment; for instance, "I hope the award will help you gain exposure so more people will know about your catering and event planning services. They really are second to none!"]. All the best,

– Your first name

Deeper Insight

There's no critical timeline on when to write the thank-you note, but you should send it close to the event or "big moment" so it has the most impact.

Remember: few people would ever take the time to send a "Congrats" email. Fewer still would send a handwritten note. Be the person who stands out. Be the one they never forget.

Thank someone who went above and beyond

Name,   Month/Day

Thank you again for [how the person went above/beyond; for instance, "rushing the order of company keychains so we had them in time for the Acme Industry Conference"]. [Then, provide detail on how the extra effort made a difference; for instance, "We passed out nearly every keychain, and it allowed us to form several key relationships with new business prospects — including two in our target market of San Diego"]. Again, thanks so much and please let me know if I can ever return the favor.

                                                  – Your first name

---

Deeper Insight

It's not enough to say, "Thanks a lot." Give the person details about exactly how much you appreciate the favor/extra effort and how it helped your business. That's why I wrote "allowed us to form several key relationships...including two in our target market of San Diego." The additional details drive home how much you value the assist.

The thank-you note should go in the mail within 24–48 hours or as you have time (in case you were traveling and not at your desk).

How to draft a letter

You know I am all about online communication, but there are several instances when you need to draft a physical letter. A few examples:

- Message to an elected official
- Announcement to clients in the physical mail
- Welcome letter in a packet for a conference

When these "formal" situations arise, how do you compose the letter? There's a standard format, which I outline below. The template is also an opportunity to create company letterhead for other printed messages.

In the example below, I provide a fictitious announcement to clients about a change in service offerings.

———

Company logo in upper left   Date in upper right

Recipient's contact information (if letter goes to one person)
[First and last name]
[Street address]
[City, State, Zip]

Dear [Mr./Ms. _____ or "Name of Company" Team],
I hope this letter finds you well, and thanks for your continued support of [name of your company; for instance, "Acme Logistics"].
[Now, the main point of the letter; for instance, "I am writing to update you and your team on a big change in the way we ship goods from the West Coast to Asia"].

> NOTE: In printed letters, avoid contractions like "I'm" or "we're."
> Spell out the words to give the letter a more polished feel.

[Then, provide detail to back up your main point; for instance, "Starting on May 9, our office will ship packages between the hours of 11 p.m. and 6 a.m. Previously, we held the packages until the next morning. We believe the additional shipping hours will help you deliver goods to Asia faster and more efficiently."]
[Finally, include any additional information; for instance, "As well, we have extended our hours at the Santa Monica processing facility for walk-in customers. We are now open until 9 p.m. Monday through Friday (originally 8 p.m.)"].
[A line to conclude the letter; for instance, "Thanks for being a valued customer, and please let me know if you have further questions"].

    – Your first and last name
     Job title, company

Deeper Insight

Make sure the letter has your logo at the top. It gives the document legitimacy and a sense of professionalism. In fact, you should have company letterhead already prepared for these and other formal situations.

Be sure to print out and review the letter for any typos. Don't scan the document on your computer; you might miss obvious errors. Printed out is the way to go.

Lastly, it's a nice touch to sign with pen next to your name at the bottom. If this letter goes out to 10,000 people, then you likely don't have the time. But if it's manageable, consider a signature and maybe even a short note unique to the recipient. Something like, "We always appreciate your business at Acme Corporation!"

The little stuff makes the biggest difference. Always.

Happy birthday/Get well

OK, I will not lecture you on how to buy and write a birthday or "Get Well Soon" card. I think we know how to do both.

The only information I will add to the topic is the importance of doing more than the basics.

Yes, if the entire team signs a "Get Well Soon" card for a client, you can write "Get Well Soon!", sign your name and be done with it. But that's the easy way out.

Take 15 seconds and think about how to make your message memorable, even if it's one of 20 well wishes on the same card.

Something like, "Feel better soon so we can get back to playing phone tag! :)"

Think about your relationship, and if appropriate, add a piece of detail that's special or significant to the other person.

Even with a routine task like signing a card, there's always an opportunity to stand out.

Awkward situations

INTRODUCTION

Email is wonderful but also has its drawbacks. Since we can't hear or see each other, our messages can sometimes confuse the recipient(s) or, at worst, offend.

In the next section, I list out several "awkward" email situations and how to avoid them or navigate your way out of them.

Keep people/company name consistent

Sometimes, we need to send the same general email to several different people, but the emails go out one person at a time.

In those moments, **be extra careful** about the person's name and, if included, the person's company. Otherwise, it's awkward to send an email to someone but include the name of the person who received your *previous* email.

For instance, here's the first email:

Hi **John**,

Good morning.

I'm writing to make you aware that our new product line of Acme power tools is now available on our website. I want to let you know because I think the products would be an added resource for the work you do at **Company ABC**.

Thanks, and let me know if you have any questions.

– Your first name

Email signature

In the next email, check your work carefully to ensure you swapped out the name and company. See the words in bold.

For instance:

Hi **Jane**,

Good morning

I'm writing to make you aware that our new product line of Acme power tools is now available on our website. I want to let you know because I think the products would be an added resource for the work you do at **Company XYZ**.

Thanks, and let me know if you have any questions.

– Your first name

Email signature

If you replaced "John" with "Jane" but forgot to switch out "Company ABC" to "Company XYZ"...awkward.

If you go the extra step and make the email to each person customized and unique (for examples, go to page 71), it's still easy to forget to update the name and company. Even if you use a mail merge, check yourself early and often.

Be careful with forwarded emails

Here's a scenario. Your team works with a client, Peter, who is overbearing and obnoxious. One morning, you and a coworker, Randall, email about the client's project and it looks something like this:

You: So you'll have the proposal ready for Peter by 2 p.m.?
Randall: Yea, I'll have it finished by then so he doesn't go crazy on us.
You: Seriously. He's a lot to deal with.

Two hours later, Randall responds to the email chain with:

Randall: Here's the proposal. You can send it over.

Without thinking, you forward the email to Peter.

But wait…uh oh. That email chain also includes the lines "so he doesn't go crazy on us" and "He's a lot to deal with."

Yikes. That's no good.

As a best practice, start new email chains when you send a message to a client. It's easy to forget what information is contained in a long thread of email responses.

Compose a brand new email, and you're in the clear.

Get on the same page

There are times when you can't discern the true meaning of someone's message, and it does no good to send *more emails*.

Here's an example:

Other person: What I'm trying to say is, the team lead wasn't upset at the presentation, but she said it could have been done a different way.

You: Ok, so she wasn't upset with how it went?

Other person: Well, she wasn't thrilled. Let's put it that way.

You: I think it would be easier if we talked this out by phone. Do you have a few minutes?

Other person: Yea, call me in five.

———

Then, on the phone you two can sort out the situation and, as the title of this section suggests, "get on the same page." That way, you avoid a complete miscommunication or the chance that you follow instructions incorrectly and do a bunch of work the wrong way.

In so many situations, email kickstarts the conversation but a phone call or in-person meeting provides clarity.

Foreign emails bound for the US

I have had conversations with foreigners who hope to do business in the US or other places where English is spoken/written.

They ask me, "I want to use email to introduce myself, but I don't have the best English writing skills. What should I do?"

You can overcome the barrier — and make the exchange less awkward — through a simple solution.

Be up front about your English writing ability.

Here's an example:

Hi Mr. Williams,

My name is [first and last name], and I am a/an [job title] at [name of company]. I'm writing to discuss a possible partnership between our two businesses. Please excuse my English as it's not my first language.

The "please excuse" line allows you to be honest and transparent. And then it will make the email recipient focus less on your words and more on your overall message.

HOWEVER, you should still do your best to make the email read well. And that also means avoid typos and misspellings as much as possible. When in doubt on a grammar rule or spelling, look it up.

I also recommend you print out your email and read it aloud to yourself. You will catch more mistakes that way rather than looking at it over and over on the screen.

My Journey: Part 3
That time I wrote my first book

Let's continue from my story on page 69. If you recall, I arrived at the idea to do a big book of email templates because my web traffic had proven the concept had potential.

OK, great. So now I'm going to write a book.

Sooo, how the hell do I write a book?

My first move was to seek the counsel of an author I met through Twitter (reason #256 why I love Twitter). I used aspects of the template on page 71 and set up a phone call.

That conversation started me in the right direction on how to assemble the manuscript. I began to flesh out the table of contents and thought, "Wow, I have so much to write!"

I had about 40 percent of the book's contents already contained on my blog so I needed to fill in the rest with brand-new email templates and other writing guides.

I found any spare minute of the day (weekends too) to chip away at the remaining 60 percent of content. It was a grind but also the most meaningful writing I had done to that point so it never felt like work. Ya know?

After about six months of writing, I had a rough draft. It was 264 pages and over 50,000 words. Whew.

The next six months began the task of laying out the pages with a designer's help. I did not want the book to feel amateur on any level. I never, ever cut corners here.

By the end of the design phase, guess how many versions I went through before I came to the final one?

Twenty three. Twenty freaking three. I'm exhausted thinking about all the iterations. But each time, the product became a little sharper, a little clearer. I made sure to use my own writing guides like the technique to edit my content in three minutes from page 9.

Then the next six months? Editing and refining. I thought the first six months (writing) were the toughest. Then I was sure it couldn't be worse than the middle six months (designing).

Oh no. Editing, revisions and looking for typos (particularly for a book on writing skills) nearly brought me to my knees.

I read the book seven times and had trusted friends and family go through it too.

I remember the day a proof of the book arrived in the mail. I couldn't wait to hold the finished product in my hands. What a cool moment — to see and feel the end result of so much hard work. I know other authors, creators and entrepreneurs understand that emotion too.

With the book created, I now turned to people who could guide me through the marketing process. In the next part of "My Journey," my mentorship "dream team" takes shape.

Chapter 4
Interoffice Communication

In the following chapter, I discuss smart ways to communicate with your own team. When you write efficiently, it helps the entire organization run faster and look more composed to the outside world.

HOW TO RECAP THE TEAM ON A BIG PROJECT

Emails to your own team carry a great deal of weight. You want the office reputation as someone who communicates effectively. That way, you save everyone time!

Here's a template to recap your team on a big project. Imagine the bold type in the email body below is yellow highlight.

Subject line: Updates on [project name; for instance, "Project Alpha"]

Hi everyone,

I have several updates about the [name of the project] so I will lay them out as clearly as I can.

[If there's a special note or update you need to include first, do so here. For instance, "We've been given strict orders from upstairs to have the first trial phase completed by March 23 so we'll need to buckle down and push through the remaining tasks."]

- [Then, list out each update or action item as a bullet point. Let people absorb every piece of information one at a time so nothing is lost in a big paragraph. Also, add a space between each bullet so they're not crammed together. Again, let the words *breathe*.

- [Whenever possible, call people out by name and put the name in yellow highlight. That way, the person will see what he/she needs to do.]

- [For instance, "**Regina** said she will continue to work on the glitch whenever we try to put the program into sleep mode."]

- ["**Donald,** can you have a updated report on the analytic capabilities by Friday?"]

- ["We can't use the visitor parking lot for testing. We'll need to find a different place if anyone has ideas."]

- ["I know we're running up against the holidays so **can everyone reply to me** with your vacation situation?"]

Thanks, and let's keep pushing to get the project done.

– Your first name

Deeper Insight

The best leaders are effective communicators. You will gain respect among team members through clear, concise emails.

How to recap a conference call or office meeting

After a phone call or meeting, what's the best way to organize the information and present it to the group via email? Again, brevity and highlighting win the day.

Subject line: Recap of [conference call/meeting/phone call] on [month and day]

Hi everyone,

Thanks for your time [on the call/in the video chat/at the meeting].

[Use this space to give everyone a short summary of the meeting; for instance, "Quick recap of the meeting: the design phase of the construction project is on track, and now we need to turn our attention to the actual build."]

These are the action items [over the next few days/over the next week/ moving forward]:

- [Then, list out each update or action item as a bullet point. Let people absorb every piece of information one at a time so nothing is lost in a big paragraph. Also, add a space between each bullet so they're not crammed together.]

- [Whenever possible, call people out by name and put the name in yellow highlight. That way, the person will see what he/she needs to do.]

- [For instance, "**Malcolm** will talk with the general contractor about his timeline and the materials needed for the brick courtyard."]

- ["**Heather** said she can create the flyer to alert all the neighbors about upcoming construction."]

- ["**Dan** and **Beth** will both work with city officials and the traffic engineer on securing the area for construction."]

- ["**Esther,** can you send everyone the Excel document with the real estate estimates?"]

Thanks, and let's keep pushing to get the project done.

– Your first name

Email signature

Deeper Insight

Send the recap email within two to four hours of the conference call or meeting so the information is fresh in your mind. You will also keep the momentum rolling on the given project.

How to update your boss on the status of a project

When you need to send a message up the chain to your boss or other leadership, here's the sharpest way to do it.

Subject line: Updates on [name of the project; for instance, "sales presentation for the conference"]

Hi _____,

Good [morning/afternoon],

I have some updates on the [name of the project] since we last spoke. Here's the latest:

- [Then, list out each update or action item as a bullet point. Let people absorb every piece of information one at a time so nothing is lost in a big paragraph. Also, add a space between each bullet so they're not crammed together.]

- [For instance, "I have all but one of the slides finished for the sales presentation. I'm waiting on additional sales data from Ari and then I can wrap everything up."]

- ["I spoke to the conference organizer, and she said we'll have 20 minutes for our presentation and 10 minutes for Q & A. Do you want me to draft a list of questions and send them to you?"]

- ["I attached the PowerPoint to this email. Take a look and let me know if it needs any further changes."]

Thanks, and I'll keep you updated when I have the presentation 100% finished.

– Your first name

Deeper Insight

The "update the boss" email is a smart move whenever it feels like a lot of stuff has happened, and the boss needs to know. If you work remotely or from home, the email is critical. How else would the boss know what you're up to?

Also, the bullet point concept is important here. Let your boss take each piece of info one at a time. Show you understand how to communicate like a pro.

How to clarify a miscommunication

When issues arise, what's the best email tactic to control the situation? Address the situation right away and don't bury the main point.

Subject line: Update on [the issue at hand]

Hi team,

I want to clarify [the issue at hand; for instance, "members of the team who will receive new computers. I know there's some confusion about who is in line for the upgrades and who isn't."]

[Then, tell people the facts and clear up the miscommunication; for instance, "Our marketing and sales teams will receive the new laptops first. They can expect the upgrades in mid-February. The rest of our team will have new laptops by early April."]

[If you need to provide extra explanation, do so here; for instance, "The reason for the staggered upgrades is because our marketing and sales teams have an increased workload over the next few weeks and need the enhanced capabilities first."]

If anyone has questions, please feel free to reach out.

Thanks,

– Your first name

Email signature

Deeper Insight

Clear up the problem right away and provide deeper explanation so you don't leave any questions hanging out there. And then offer to answer further questions as they arise.

How to welcome a new employee to the team

If you're in a leadership role, it may fall to you to introduce a new team member. Certainly, you'll want to show him/her around the office. But it might also make sense to email the team ahead of time.

Subject line: Meet [employee's first and last name], our new [job title]

Hi everyone,

Good morning/afternoon.

I want to introduce [employee's first and last name]. He/she is a new [job title] and will [describe in one line the person's primary role]. We're excited to bring [employee's first name] on board and add him/her to the mix.

116

[Then provide brief background info on the new hire; for instance, "Mark graduated last fall from Big State University with a degree in political science. He spent a semester as an intern for Senator John Doe so he has experience on Capitol Hill."]

[Then, include when the person will start in the office; for instance, "Mark will begin on Monday, May 2 in our downtown DC office."] Please make sure to welcome him/her to the team.

Thanks,

– Your first name

Email signature

Deeper Insight

First and foremost, give your employees a heads up that someone new has joined the company/organization. Then, shed some light on the person's background and start date so your team knows more than his/her name and job title.

My Journey: Part 4
That time I built my mentorship dream team

Many of us look at the concept of a "mentor" the wrong way. It's unlikely you'll find one person who can address every career challenge or obstacle. No one is an expert at *everything*. Plus, if you rely on one mentor all the time, you will wear the person out.

The better approach is to assemble your mentorship "dream team" or perhaps a personal "board of advisers." Each mentor has expertise on a given topic, and you can ask for his/her specific guidance when necessary.

To write/edit/publish my first book and continue to develop my brand as a communications specialist, I looked to my mentorship "dream team" for advice and direction.

When I ran into a wall (still run into walls) with book publishing, I reached out to two people who understand the process — one who edits book manuscripts and another who designs books covers/interiors. I often used elements from my email template on page 71 (how to ask for advice) to start the conversation.

As I explored (still explore) ways to deliver in-person workshops and presentations, I booked time with more seasoned professionals and learned their methods.

And as I tried (still try) to understand the nuances of digital marketing and audience building, I asked (still ask...OK, you get the picture) blogger friends to review my efforts and share what works best for them.

In short, I built a dream team and gathered powerful insights from all sides.

My dream team grows to this day, and I plan to keep everyone "on my board," as long as they'll have me. There's always another challenge up ahead and, fortunately, someone with the knowledge to help me through it.

Speaking of which, my dream team came together at the right time. I had big plans for my shiny, new book but needed clear instruction on how to promote it. I sought the counsel of my mentors, received solid advice and then rolled up my sleeves as a one-man PR machine.

In the next section of "My Journey," the book launch begins.

Chapter 5
LinkedIn Writing Guides

Note: LinkedIn often changes its features and how information is displayed. You may need to amend my advice in this chapter if you find it doesn't match with LinkedIn's current functionality.

WHY YOU SHOULD PUBLISH YOUR OWN CONTENT

LinkedIn offers business professionals the world over a terrific PR opportunity.

When you publish content on LinkedIn, the information is then shared with people in your own network AND has a chance to reach an even broader audience if LinkedIn chooses to promote it further.

In a sense, you then have a column appear in a "digital newspaper" that's read by all the people connected to you in the business world. It's an excellent opportunity to prove your expertise with minimal effort and zero cost.

Log into LinkedIn and click "Write an article." The instructions are pretty straightforward from there.

DO NOT grab an image from Google and use it for your article. The image must be legal to obtain (a great place to look for fair-use photos is Flickr.com). It's also wise to attribute the image via the text in the upper left corner of the image box.

Here's the beauty of a LinkedIn post: it can be the same content that you posted on your company website. Why not use it twice and reach more eyeballs?

For blog post ideas, go to page 165. There, I provide topics for several broad categories.

Start writing and then start repurposing content on LinkedIn!

How to build key relationships on social media

Meaningful relationships take time. The same is true on social media.

You can't, for instance, use Twitter to run up on people you respect and bombard them with pitches about your products and services. **To gain trust, you must play the long game.**

"But Danny…the long game? Ugh, I don't have time for that. I need results TODAY."

I know. I get that. We all want success ASAP. But if you hope to secure relationships that last, you need to invest in other people and show them you're interested in what they do.

> Note: if you want the attention of a "famous" person with hundreds of thousands (or millions) of fans, I can't guarantee this "give before you get" approach will work. But here's what I know to be true: the "famous" person will more likely respond to you after you've shown how much you value him/her. A one-time "look at me!" attempt won't work.

Here are four ways I have built relationships via social media with others I respect in the career advice/professional development space. The people I befriended have boosted my career in countless ways, and I try to support them too. But let me be clear: for someone to feel I was worth their time, I had to prove my commitment over weeks, months and sometimes even years.

Here's what to do:

1. Share other people's work — a lot

Maintain a list of websites from bloggers and other influencers you follow and respect. Each week or as often as you can, share their newest content on your own social media channels and tag the person/company. The blogger/influencer will notice and over time come to respect your selflessness. Perhaps at some point, the person will even respond to your post and you can start a conversation. And from there, a relationship begins to form. Remember: always ask how **you** can be a resource and not what the person can do for you.

Pro tip: I spend time on Sundays scheduling posts throughout the week in which I share other people's content. That way, the work is done once Monday rolls around and life becomes hectic.

2. Comment on other people's work — a lot

Don't get me wrong — it's smart to share other people's work on social media week after week. But it's even BETTER if you also comment on the post through the person's website. Comments are hard to come by, and most bloggers/influencers relish new comments and enjoy responding to them. You don't need to chime in on every single post the person writes, but instead make "blog commenting" an ongoing strategy and do so when you can. The tactic will help you develop trust and build on the relationship.

3. Offer to let the person guest post on your site

We often only think of "guest posting" as a way to spread our own message on someone else's platform (see page 61 to learn how to ask to submit one). But let's flip the idea: ask if the blogger/influencer wants to guest post on your site. OR see if the person wants to provide an expert quote on a topic you plan to write about. OR ask if the person wants to be featured in a Q&A on your site. The point is to look for ways to give the person the spotlight (to grow his/her business).

4. Rome wasn't built in a day

Big stuff takes time, and so often our success depends on the strength of our network. Few people will dive in headfirst until they "know" you. That means they have to trust you're for real and not some shady character coming around to trip them up.

Play the long game. Invest in relationships. THAT'S how you build a solid network.

Memorable business and personal profiles

PROFESSIONAL HEADLINE

On LinkedIn, we can share our personal brands with the world. Can you guess where?

In the professional headline, the space right below your name.

Most people use the line to write their job title.

John Doe

Project Manager at Acme Corporation

Sure, that's appropriate and won't get you in trouble. But here's the catch: *most people list their job title and company, which makes their LinkedIn profiles blend in with all the rest.*

Also, *Project Manager at Acme Corporation* isn't a professional headline. It's just the facts as if to say, "This is what I do, and this is where I work."

OK. But what's your brand?

Maybe John Doe excels at data analytics, and he's become known around the office for his ability. Then his professional headline could be:

John Doe

Using data to make smarter decisions

Or perhaps:

John Doe

Powerful insights driven by data

Yes, your job title and company matter, but your "brand" is more interesting. It might catch readers by surprise and lure them into your profile.

So how do we craft a professional headline?

First, ask yourself this question: *where do you provide the most value on the job?*

If you're in customer service, then the headline could be "The customer always comes first" or "Dedicated customer service specialist."

If you work in IT, the headline could read "Ready to solve the toughest tech challenges" or "Cybersecurity and antivirus expert."

Think about how your skills allow you to make an impact on others. Why do you matter? Then turn the answer into a short phrase.

That's your professional headline. That's your brand.

A few more points to consider

- Don't use the professional headline to brag. For example, "Greatest marketer in the country." Nope, probably not. Instead, tell us how you make others better.

- Please don't write the exact phrase "Turns complex problems into solutions." It's cliché and overdone.

- Keep the professional headline to eight words or fewer. Otherwise, it will drag on.

- Ask a few friends or co-workers what they think of your headline. Tell them to be honest and not hold back.

- Once you set the professional headline, forget about it for a couple of hours and then look again. Do you still like the headline or does it feel funny? Listen to your gut — it's usually right.

Profile summary

You have 30 seconds to describe yourself. Can you do it?

With a LinkedIn summary, that's all the time you have. Sorry, no one wants to read your entire work history. Not even a little bit.

The summary section requires brevity and critical thinking. You must explain what you're about and the impact you make on others. As well, the template I provide allows you to seem approachable and authentic. Let your writing make you *human* rather than feel like a bunch of corporate buzzwords and jargon.

Let's begin with step 1, and you'll see what I mean.

In my first book, *Wait, How Do I Write This Email?*, I include a fictitious LinkedIn profile summary for an IT professional named Lamar.

Here, I include a new profile summary for a traffic engineer named Sheila.

Step 1: Who are you, really?

Keep it basic. In a nutshell, what are you known for? What's your identity? And how does your work help other people?

It's a useful exercise to describe yourself in less than ten seconds. Plus, it's important to think hard about the value you add to the market.

"I travel the state of Louisiana to monitor traffic patterns in congested areas and ensure drivers can use the roads safely."

At Sheila's professional core, that's the work she does AND how her work improves the lives of other people.

Step 2: What do you do?

Now take the opening line a bit deeper, but remember the 30-second rule. This is no time to delve into three huge paragraphs on everything you've done. Keep it rolling with specific details, like:

- Your title and company

- BRIEFLY what you do at the job

- Again, how your job *helps people*

- Specialties, areas of expertise and statistics to describe your role a bit further

"As a traffic engineer for Acme Engineering Corporation, I log 10,000+ miles a year crisscrossing the state to evaluate 200+ busy highways and biways. I specialize in finding ways to integrate bicycle lanes into existing city streets as well as intelligent transportation systems that direct traffic in urban areas. I keep a close watch on traffic engineering globally and in the US so I can bring best practices to Louisiana."

Step 3: Bring 'em home

In the final step, put a stamp on your LinkedIn summary.

By now the reader knows who you are and what you do. Now, finish out with a strong "closer" sentence.

Similar to your opening line, what's your mission as a working professional?

What are you passionate about? And I know I'm a broken record but...how does your passion help other people?

Reiterate that point at the end, add a period and you're done.

Sheila's closing line:

"When I watch drivers in my state travel without a hitch, I feel proud and know I did my part to keep Louisianans moving ahead."

Sheila's full LinkedIn profile summary

I travel the state of Louisiana to monitor traffic patterns in congested areas and ensure drivers can use the roads safely.

As a traffic engineer for Acme Engineering Corporation, I log 10,000+ miles a year crisscrossing the state to evaluate 200+ busy highways and biways. I specialize in finding ways to integrate bicycle lanes into existing city streets as well as intelligent transportation systems that direct traffic in urban areas. I keep a close watch on traffic engineering globally and in the US so I can bring best practices to Louisiana.

When I watch drivers in my state travel without a hitch, I feel proud and know I did my part to keep Louisianans moving ahead.

———

In three short sections, we gain a deep understanding of Sheila and why she's passionate about her job. She also provides statistics ("10,000+ miles" and "200 busy highways and biways") so we understand the scope of her work.

Finally, at the end Sheila "brings it home" with a line that says: this is my job, this is what I love to do and this is what it's all about.

There are a lot of LinkedIn profiles out there, but Sheila's summary section is one you will remember.

Why your profile summary needs to be in the first person

How many of us write our LinkedIn profile summaries in the third person? I have rewritten many LinkedIn profiles for clients, and most of the time I need to switch the copy from third person to first person.

Why? Because in a business context, we think it should feel as though someone else talks about us. The reality: we're writing about ourselves.

Plus, third person feels impersonal. There's no emotion or passion to your words. It's like you're penning a book report about yourself, and the tone is flat and forgettable.

Ah, but first person. Now we're talking. With a steady dose of "I," "me" and "my," the LinkedIn profile comes to life and allows you to be, well, you.

> NOTE: In the Experience section on LinkedIn, don't use first or third person at all. Start each line with active verbs. Instead of:

- *I manage a team of 17 people in the company's San Antonio regional office*

The line is written as:

- *Manage a team of 17 people in the company's San Antonio regional office*

The LinkedIn profile summary requires a subject, and that's where I advocate first person over third.

In my first book, *Wait, How Do I Write This Email?*, I provide an example of a LinkedIn profile summary for a fictitious IT professional named "Lamar." Here's the opening line from page 167:

Every day, I protect sensitive information on thousands of people from hackers and cyberattacks.

Now read the same line in the third person.

Every day, Lamar protects sensitive information on thousands of people from hackers and cyberattacks.

The second version sounds, again, like someone wrote a report *about* Lamar. It doesn't feel like Lamar is accessible and approachable.

You might wonder, "Hang on, it's OK to use first person on LinkedIn? That doesn't seem professional."

Not only is first person OK, but I believe it's **essential**. LinkedIn is a place to network and build relationships. Why would you talk in the third person and act like you're not "there" to mix and mingle?

The word "I" suggests you're the one behind the keys ready to engage in conversation. The approach makes you genuine, authentic and **real**.

And isn't that how you want people to think of you?

Experience section

The "Experience" section is your chance to show how much you accomplished on the job (or during internships/school).

There are several strategies to improve the bullet points under your work experience. I will use a fictitious job to describe each one.

Here's the current version before we improve it.

Acme Corporation — Austin, Texas — March 2014-present

Assistant Manager

- Manage warehouse facility and oversee DSD inventory process
- Track incoming and outgoing shipments to ensure on-time delivery
- Won company award in 2015 for "Best Managed Warehouse"
- Facilitate trainings for new hires as well as continuing education for all other employees

Note: If the bullet point does not contain a complete sentence, leave out the period.

OK, on the surface, you might think the bullet points look fine. But when we go deeper, you will see many missed opportunities to stand out.

Let's go through each bullet and make it better.

- Manage warehouse facility and oversee DSD inventory process

Rule #1 with business communication: never assume the person knows what you're talking about. **Ever.**

First, what products are contained in the warehouse? Where is it located? How big is it? We have no idea. Also, what's "DSD"? No clue.

The first bullet point should provide context so people understand the nature of the work. And we should always spell out abbreviations on first reference. If the abbreviation also needs a bit of explanation, provide it too.

The corrected bullet point:

► Manage 200,000 square foot warehouse facility in Detroit that stores home appliances like microwaves, refrigerators and ovens; oversee our Direct Shipping and Delivery (DSD) inventory process, robotic technology that takes items into and out of the facility

Now, anyone in the world who reads the LinkedIn profile will understand what this person does every day. Plus, the first bullet contains key details so it's memorable (ex: 200,000 square feet, Detroit, microwaves, robotic technology).

All right, moving on to bullet #2:

► Track incoming and outgoing shipments to ensure on-time delivery

How can we make this bullet point jump off the page? One word: quantify. Numbers make the work appear more impressive. How many? How often?

► In a given week, track 34,000 incoming and outgoing shipments to ensure on-time delivery (over 1.7 million shipments a year)

Wow — 1.7 million! Do you see how much stronger the bullet point becomes when you quantify your efforts?

Third bullet:

► Won company award in 2015 for "Best Managed Warehouse"

OK, this person won an award. That's great but again, where's the context? How many people did the person beat out? How many warehouses does the company have? Quantify quantify quantify!

► In 2015, competed against 93 coworkers at 26 warehouses to win company award for "Best Managed Warehouse"

See the difference? Now we know how impressive it is to win the award. This person beat out 93 coworkers. That's a lot!

Last bullet point:

▶ Facilitate trainings for new hires as well as continuing education for all other employees

Again, how many people receive the training? The line has no "umph" until you add numbers.

▶ Facilitate trainings for about 130 new hires every year as well as continuing education for 400+ existing employees

One more point: storytelling is a dynamic strategy that can elevate every writing opportunity, even a LinkedIn "Experience" section. Here's an example of a mini-story inside the final bullet point about employee training.

▶ Facilitate trainings for about 130 new hires every year as well as continuing education for 400+ existing employees; once led a training on emergency evacuations as the fire alarm went off accidentally (took command and led over 200 people safely outside)

Quick story. Memorable addition to the profile.
Here are the bullet points "before" and "after."

Acme Corporation — Austin, Texas — March 2014-present
Assistant Manager

▶ Manage warehouse facility and oversee DSD inventory process

▶ Track incoming and outgoing shipments to ensure on-time delivery

▶ Won company award in 2015 for "Best Managed Warehouse"

▶ Facilitate trainings for new hires as well as continuing education for all other employees

Acme Corporation — Austin, Texas — March 2014-present
Assistant Manager

- ▸ Manage 200,000 square foot warehouse facility in Detroit that stores home appliances like microwaves, refrigerators and ovens; oversee our Direct Shipping and Delivery (DSD) inventory process, robotic technology that takes items into and out of the facility
- ▸ In a given week, track 34,000 incoming and outgoing shipments to ensure on-time delivery (over 1.7 million shipments a year)
- ▸ In 2015, competed against 93 coworkers at 26 warehouses to win company award for "Best Managed Warehouse"
- ▸ Facilitate trainings for about 130 new hires every year as well as continuing education for 400+ existing employees; once led a training on emergency evacuations as the fire alarm went off accidentally (took command and led over 200 people safely outside)

Which person do you find more experienced (and interesting)?
I thought so.

Company page description

Like a personal profile summary, LinkedIn allows us to write a description for our businesses on a company page.

We should use the same three-step method I explained on page 122 (personal profile summary).

Step 1: Who are you, really? — "nuts and bolts" description of your company and the work it does

Step 2: What do you do? — deeper explanation on your products/services

Step 3: Bring 'em home — why your business matters and the impact you make on the world

As an example, here's the company page description for a fictitious business known as Acme Wealth Management.

Step 1: Who are you, really?

Acme Wealth Management is a team of experienced advisers that helps clients in the Portland area achieve their financial dreams.

> *NOTE: In one clear line, this is who we are and what we do.*

Step 2: What do you do?

At Acme, we put your needs first and build an investment plan that's designed to evolve as your life changes. We specialize in retirement planning and strategies necessary during significant life events like the loss of a loved one, divorce or inheritance. We follow the latest industry rules and regulations to ensure we provide sound judgment to every client. And we have been named among "Portland's Best Wealth Managers" by *The Portland Daily Newspaper* for the past five years.

> *NOTE: Above, the company provides a deeper explanation with details like specialties (ex: retirement planning, divorce) and honors (ex: "Portland's Best Wealth Managers").*

Step 3: Bring 'em home

We believe every client presents unique opportunities to build wealth that will last a lifetime. Contact our office today to start the conversation.

> *NOTE: Closing line that sums up why you're passionate about the work as well as a call to action.*

Here's the entire company page description:

Acme Wealth Management is a team of experienced advisers that helps clients in the Portland area achieve their financial dreams.

At Acme, we put your needs first and build an investment plan that's designed to evolve as your life changes. We specialize in retirement planning and strategies necessary during significant life events like the loss of a loved one, divorce or inheritance. We follow the latest industry rules and regulations to ensure we provide sound judgment to every client. And we have been named among "Portland's Best Wealth Managers" by *The Portland Daily Newspaper* for the past five years.

We believe every client presents unique opportunities to build wealth that will last a lifetime. Contact our office today to start the conversation.

Outreach templates

WHY YOU SHOULD AVOID DEFAULT MESSAGES

When you connect with people on LinkedIn, it asks you to customize the invitation. **Do it.**

Why should you add a special note? Three reasons:

- ▶ If you write an authentic message, the person is more likely to respond.
- ▶ You can use the message to start a conversation and build on the relationship.
- ▶ People won't expect to see a message, and the move will help you stand out right away.

Here are examples of different business scenarios.

If you meet someone at a networking event:
Great to talk with you at the Chamber of Commerce happy hour. I enjoyed learning more about your landscaping business.
– Your first and last name

If you find the person's profile or website interesting:
I enjoyed reading your blog, Acme Tech Talk. Your advice helped me troubleshoot a problem with my new tablet. Thanks for the help!
– Your first and last name

If you might like to work together:

I work at Acme Automotive and we are looking for a new advertising agency. Please let me know if we can talk further over email or phone.

– Your first and last name

To increase your odds at a response, ask a question. Here's an update to the "networking" message from up above:

Great to talk with you at the Chamber of Commerce happy hour. I enjoyed learning more about your landscaping business. Remind me: did you say you handle corporate office parks?

– Your first and last name

There are so many reasons why you would "connect" with someone on LinkedIn. No matter the situation, remember to make the outreach personal. Few people will respond to the standard, cookie-cutter message. Stand out and do your best to start a dialogue.

How to write a networking message

LinkedIn has made private messages more like a back-and-forth conversation. If you press "Enter" to start a new line, your text is automatically sent.

That's why you should first uncheck the box within your message area that says "press enter to send."

Then, you can compose your message and not worry if it will send too soon.

Hi _____,

My name is [first and last name], and I'm [put yourself in context; for instance, "the owner of Acme Party Supplies, the largest store of its kind in the Orlando area: add URL here"].

I'm reaching out because [explain what you want in one or two sentences; for instance, "I see your company, Acme Party Plus, specializes in clowns and face painters for birthday parties. I think there may be ways for us to collaborate"].

NOTE: Put your "purpose" at the top of the message so the person sees it right away.

[If you have more information about the request, include it here and consider bullet points to make everything easy to read.]

[Then, reference a couple of facts from the person's LinkedIn profile to prove you studied up before you sent the message; for instance, "I read your LinkedIn profile and see you've provided face painting since 1997. I also noticed you do graphic design, and I might need those services too for a new brochure."]

NOTE: The person puts his/her bio right at your fingertips. There's no excuse if you leave out a section on why you find the person's career compelling.

Please let me know if there's a time over the next week for us to talk.

Thanks, and I hope to hear from you.

– Your first and last name

Deeper Insight

State your purpose at the top of the message and reference details from the person's LinkedIn profile. Make the note authentic (even though the person might be a stranger) to build trust and give yourself a better chance at a reply.

How to network with someone who sent you a connection

If someone you don't know asks to connect on LinkedIn — and you think the person is worth a further conversation — send back a message, be curious and explore. What's the point of LinkedIn if not to meet people who cross our paths?

Hi _____,

I'm [first and last name], a/an [job title] at [company if you're employed] in [city]. Nice to meet you.

NOTE: Provide a full introduction rather than only your name.

Thanks for connecting with me. I enjoyed reading more about your career, especially [one item you find interesting and a question about it; for instance, "how you and your two college friends founded Acme Startup and recently hit 100,000 downloads. What's your goal for the coming year?"].

I'm not sure if you looked over my profile, but you might be interested in my [what you want the person to know; for instance, "own project called ABC Startup, a mobile delivery service for office supplies: add URL here"].

[Then, what do you want from the other person? For instance: "Could I ask you a few questions by phone about the Columbus tech scene? I'm still new in town and want to be plugged in better. I feel like you could provide solid guidance."]

NOTE: In this case, I put the "ask" (a phone call) at the bottom because the message flows better if you first explain why you find the person's career notable. You need to clarify why you reached out in the first place.

Thanks,

– Your first and last name

Deeper Insight

If someone connects on LinkedIn with YOU, then why not send a quick message, network and see if there are ways to work together or collaborate?

If you receive no response but feel it's a relationship you need to pursue, wait 48 hours and reply back with, "Hi _____, Please let me know you saw my message from [day of the week]. It would be great to connect and talk further. Thanks."

How to network with someone after he/she accepts your connection

If you make the initial outreach but don't know the person, consider a message like the one below after the person agrees to connect.

Hi _____,

I'm [first and last name], a/an [job title] at [name of company if you're employed] in [city]. Nice to meet you.

Thanks for accepting my connection request. I enjoyed learning more about your career, especially [one tidbit you find interesting and a question about it; for instance, "how you switched from a career in communications to the medical field. And congrats on your recent grant to study the effects of public drinking water on children in public schools. When do you expect to complete your study?"].

I'm not sure if you looked over my profile, but you might be interested to know [what you want the person to know; for instance, "I am in the communications field as a local news producer. Here's a quick example of my work: add URL here"].

[Then, what do you want from the other person? Perhaps you can use a line like, "I am also thinking about a move to the medical field. I enjoy writing news stories on science the most and want to explore the industry further, particularly biomedical research. Would you have time for a quick phone call?"]

NOTE: In this case, I put the "ask" (a phone call) at the bottom because the message flows better if you first explain why you find the person's career notable. You need to clarify why you chose to connect in the first place.

Please let me know what's possible.

Thanks,

– Your first and last name

Deeper Insight

Since you initiated the relationship with the invitation to connect, don't be too aggressive with the follow-up message. Start with proof you explored the other person's career and finish with a call to action (phone call, meeting, etc...). If you receive no answer, refer to the "Deeper Insight" on page 136.

How to ask someone to give you a recommendation

What makes a great LinkedIn recommendation? Stories, stories, stories. Tell the person to skip all the empty rhetoric about how "amazing" you are. Rely on your experiences and how you delivered in the clutch.

Hi _____,

I hope you're doing well. [A quick line to make conversation; for instance, "Has your team finished the big move to the new office? I hope it's going smoothly."]

I'd like to add a recommendation to my LinkedIn profile about [how the person would praise you; for instance, "how our company helped your team create new business cards and letterhead in advance of the move"]. Can you write a review for me? It can be brief like three or four sentences.

137

[If you and the person share a memorable story that makes you look good, ask the person to write about it. For instance, "If you want, you can write about how our team received the final artwork 36 hours before you needed the materials printed and we still finished the job on time."]

> NOTE: Direct the person towards a fitting story. He/she might pick another but at least you offered an example to make it easier. If you don't have a clear story in mind, let the person know what you want from the recommendation. For instance, "I'd like the recommendation to show other people our team is reputable and works hard to do the job right."

If you can write the review, I'll send you a LinkedIn message with the request. I'm also happy to write one for you.

> NOTE: Ask first for the recommendation and then send over the actual request through LinkedIn.

Please let me know and thanks in advance,

– Your first name

Deeper Insight

Ask if the person would also like a recommendation on his/her profile and return the favor. And once the person does write a recommendation, be sure to send another LinkedIn message to say thanks.

How to network with alumni

Alumni from your school (high school, college or any other "school" or "program" you completed) are ideal people to contact for networking purposes.

Why? You and the person share common experiences, which means you already have the beginnings of a relationship before you ever communicate.

Hi _____,

I'm [first and last name], a [student/alumnus/alumna] from [name of school]. I hope you're doing well.

[Start out with a sentence that speaks to your shared alma mater; for instance, "I see at Tech University you were in the marching band and played trombone. I played in the band too and was the drum major my senior year. I loved being in the band all four years."]

I'm reaching out because [Why are you sending the message? For instance, "I am starting my own interior design business, and I see you have decades of experience in residential home building. I hope you can give me some pointers as I start out and perhaps connect me to the builder community here in Tacoma"].

Please let me know if you have a few minutes this week for a phone call. I would appreciate the chance to ask questions and learn more.

Thanks so much,

- Your first and last name

Deeper Insight

The best way to land a response is to ask for advice and then talk about your shared experience at the same school. If you can't find common ground (ex: "marching band"), consider a more generic approach like, "It's always great to find Tech University alumni here in Tacoma. We need to stick together!".

Then, ask for a phone call rather than a further discussion over email. It may be too intrusive to push for an in-person meeting right away. After the phone call, a great next step is to meet in person to take the relationship further.

How to write someone you met through a LinkedIn group

If you encounter someone interesting in a LinkedIn group or through a mutual friend/colleague, send an invitation to connect. If the person accepts, follow up with a message to become better acquainted.

Hi _____,

I'm [first and last name], a/an [job title] at [name of company if you're employed] in [city]. Nice to meet you.

Thanks for accepting my connection request. I enjoyed learning more about your career, especially [one tidbit you find interesting and a question about it; for instance, "how you've grown your family's hardware store into a franchise with 15 locations across Oklahoma. Do you have plans for more stores in the coming years?"].

I'm not sure if you looked over my profile, but you might be interested in [what you want the person to know; for instance, "my company, Acme Landscapes, and our commercial landscaping services. Here's a photo of a recent shopping center we completed: add URL here"].

NOTE: Reason #347 why you need a blog or portfolio. There are so
many opportunities in networking correspondence to link people
to your work so they can see your ability on display.

[Then, what do you want from the other person? For instance, "I'd like to discuss Acme Landscapes further and see if we could provide any services. Please let me know what you think, and if we can chat by phone when it's convenient for you."]

NOTE: Once again, I put the "ask" (a phone call) at the bottom
because the message flows better if you first explain why you find
the person's career notable. You need to clarify why you chose to
connect in the first place.

Thanks,

– Your first and last name

Deeper Insight

Don't overthink the networking game. If you find someone who could become a useful contact, send the message and see what comes back.

If you don't receive an answer after 48 hours, send a short follow up like,

"Hi _____, Please let me know you saw my message from [day of the week]. It would be great to connect and talk further. Thanks."

My Journey: Part 5
That time I became a one-man PR machine

I thought writing *Wait, How Do I Write This Email?* would be the toughest act of my career. It took nearly everything I had to make those 250+ pages a reality.

Once I decided to publish on my own and rely on myself for all the marketing, I faced another uphill climb.

Wait, how do I sell a book?

I read some blogs on book marketing and learned what I could. But I recognized there were no shortcuts. I needed to hit the phones and jump on email to spread the word.

In short, I became a one-man PR machine. I sent countless emails to schools and organizations about my book (they often took the shape of a press release as described on page 34).

I used elements of the email to introduce myself to a company for the first time (page 71). And I sent *plenty* of messages to ask if I could be interviewed on a podcast (page 57) or submit a guest post that, in subtle ways, would promote my book (page 61).

Oh, and lest I fail to mention the dozens of trips to the post office to send physical copies to key people. Did you find my books in a retail location? That only happened because I mailed one of those physical copies to a potential distributor which (somehow) agreed to take me on. Total shot in the dark, trust me.

When I say one-man PR machine, I mean it. Every day. No breaks.

I believed in the potential of *Wait, How Do I Write This Email?* but also understood the book's success depended (still depends) on my promotional efforts.

And soon enough, I would learn the true meaning of the words "follow up." The hard way.

Chapter 6
Website Content

HAVE A CONVERSATION WITH THE READER

Let's talk about writing for your website.

OK, stop. See what I did in the first line? The word choice is casual, friendly and uncomplicated. The line feels like I'm talking *to you*, not writing *at you*.

On your website, there's a clear difference between content that feels stuffy and corporate versus language that's more conversational.

You might think, "But my business *is* corporate, and we need to retain a level of professionalism." Right, I get that. As an example, I have written website content for several law firm websites and, in those situations, the pages can't become too casual. It's a suit-and-tie environment, after all.

But there's still an opportunity to make the words feel like you're talking *to* the readers rather than *at* or *over* them. As an example, here's a line about how seriously the law firm takes a client's case.

Stuffy, corporate version: At Acme Law Group, we understand the importance of fighting on your behalf to get the justice you deserve. For over three decades, we have provided uncompromised excellence in the courtroom for a wide variety of cases.

The "stuffy" version is passable — no one would call and complain — but the sentences feel impersonal and a bit distant.

Conversational version: At Acme Law Group, we know it can be stressful to fight a legal battle alone. That's why, for over three decades, we have stood by our clients and worked hard to win their cases.

Does the "conversational" version feel differently to you? Instead of "we understand the importance of fighting on your behalf," it now reads, "we know it can be stressful to fight a legal battle alone." The second version is something a lawyer might actually say to a client. It's a real, human interaction.

Again, the first version: "For over three decades, we have provided uncompromised excellence..." A little too corporate for my liking. The second version: "That's why, for over three decades, we have stood by our clients..." It's more personable but still professional and worthy of a law firm website.

A "write like you talk" strategy helps to show authenticity and build trust with the reader.

Common pages

MISSION STATEMENT

A company's mission statement is a classic situation where we feel the need to use stuffy, corporate language.

We think, "OK, this is our *mission*. We have to go big here and blow people away with a line that shows we're the greatest ever."

That's when we end up with a mission statement like this:

> *"At Acme Corporation, our mission is the pursuit of quality and excellence in all its forms."*

It feels like a grab bag of corporate jargon mashed together to form a single sentence — "pursuit of" and "all its forms." Who talks like that? No one.

Also, it's unclear how Acme Corporation helps people. What problem does it solve?

A better mission statement accomplishes two goals:

1. Feels like something a person would actually say
2. Explains how you solve a problem

Here's an example for Acme Corporation, which provides e-newsletter software for small businesses.

> *Our mission is to help you write better e-newsletters so you can grow your audience and build a successful business.*

Doesn't it feel like the writer talks *with* us (not *at* us) in a conversational tone? Sections like "help you write better e-newsletters" and "so you can grow your audience" are phrases people say to each other in real life. They're authentic and believable.

Plus, the mission statement explains how the product/service solves a problem *and* makes other people better. In one sentence, we covered both goals.

When it's time to compose your own company mission statement, ask yourself:

- ▶ Do I talk *with* the audience or *at* the audience? You want a tone that suggests "I'm right here with you, and we're in this thing together."

- ▶ Do I explain how I provide a solution? Make sure to demonstrate value.

- ▶ Am I speaking from the heart? Trust me, the audience can tell. To put yourself in that frame of mind, do the following exercise.

Imagine you are face to face with a customer and have one sentence to explain why your company matters to him/her. What would you say?

The answer is the foundation of your mission statement and a great place to start as you write it down for the website.

About us

With the "About Us" page, I break the topic into two categories:

1. Short "who we are" description
2. Longer "company history" description

In the following section, I provide templates and my thoughts on both.

145

Short description

On this page, you want to give the reader a quick overview of your company, what you offer and why you matter. If the page has four or five huge paragraphs, you will lose the person's attention.

Short and sweet is the way to go.

You will see the template has the same structure as the LinkedIn company page description from page 130. Why? The website and LinkedIn company page both require brevity and a description of why your business matters.

Step 1: Who are you, really? — "nuts and bolts" description of your company and the work it does

Step 2: What do you do? — deeper explanation of your products/services

Step 3: Bring 'em home — why your business matters and the impact you make on the world

Here's the "About Us" short description for the Acme Corporation website, which sells used sports equipment in the Tulsa area.

Step 1: Who are you, really?

At Acme Corporation, we have the largest selection of used sports equipment in the Tulsa area so you can keep an active lifestyle.

NOTE: In one clear line, this is who we are and the value we add.

Step 2: What do you do?

Since 1987, we have provided high-quality gear so people can stay active and play the games they love. We have seven locations across Tulsa including our 22,000 square foot superstore in Downtown. Basketballs, soccer balls, lacrosse sticks — even complete disc golf sets. You need it, we got it. If you're looking for a unique item, visit our <u>online product directory</u> or give us a call.

NOTE: Deeper explanation with details like the number of store locations, size of the superstore and types of equipment it sells (ex: "even complete disc golf sets"). Be specific when you describe your business. Vague gets you nowhere.

Step 3: Bring 'em home

At Acme, we believe sports are central to a healthy lifestyle. Get out and play!

NOTE: Closing line that sums up why you're passionate about the work and enjoy helping others.

Deeper Insight

You might think, "That's not a very long 'About Us' section." You're right — it's not. People are busy and only need the quick summary on an "About Us" page.

If you explain who you are, what you do and why you matter, you cover all of your bases in a section of content that takes about 30 seconds to read.

Company history

You may decide to include a page with a company history to explain "how it all began."

I think an "Our History" page is a smart addition because it offers a deeper understanding of your business. If people want the quick overview, they can always read the short "About Us" page. You're covered either way.

Here's the rundown for the "Our History" page:

Part 1: A compelling story about how the company came to be.

Part 2: Details about the company's growth through the years.

Part 3: A section that describes "where we're headed next."

Even though it's tempting to tell the *complete company history,* remember to respect the reader's time. If you want the person to read the entire page, I recommend a length of 400–600 words. Beyond that, you run the risk of losing the person's attention.

Here's a fictitious "Our History" page for Acme Corporation, one of the largest car dealers in Texas. I could have also written the "Our History" section in the first person (ex: in the voice of the company's founder). That approach will be effective if it feels appropriate for your team.

———

[PART 1]

Wade Miller never thought he would own a car in his lifetime much less sell them.

The youngest of nine children, Wade was born in 1937 and grew up in poverty in East Texas. His father didn't purchase a car for the family until 1951 when Wade was 14 years old. But once the car pulled up in front of the house, Wade fell in love. He would volunteer to wash the Acme Roadmaster 1940 every weekend so his father would take him on rides and even let him grab the wheel.

When Wade turned 18, his father gave him the Roadmaster. From that day on, Wade vowed he would make cars his lifelong passion. With a high school diploma and $61 in his bank account, he set out to realize his dream.

[PART 2]

For the first year after high school, Wade worked as a land surveyor for Acme County. To save money, he lived with an older brother, Hal.

Then in August 1956, Wade attended a land auction and placed a bid on five acres well outside the downtown area; a part of town no one much cared for. Wade won the auction and purchased the land for $356. Right away, he knew what he would do: open his own car dealership.

Wade called the company Acme Corporation. He started by selling used cars purchased from auctions across the county. Eight years later, in 1964, Wade had built the largest used car dealership in East Texas. The company averaged 1,200 car sales a year and business was booming.

In the mid 1960s, Wade decided it was time to sell new cars too. Over the next two decades, Acme Corporation grew into a car dealer powerhouse with 17 locations and volume that exceeded his three main competitors combined.

By the 1990s, Acme Corporation was a household name in East Texas, and everyone knew the commercial in which Wade rode around town with his dog, Rex, sitting in the passenger seat. And as the company entered the 21st century, it remained one of the most trusted brands in the region, regularly besting his competitors in total volume.

[PART 3]

Today, Wade is in his early 80s and "semi-retired" from the business, as he likes to say. His son, Wade Jr., and daughter, Julie, run the business, but they know Wade is always there to provide advice.

Wade Jr. and Julie now have plans to expand Acme Corporation into other parts of Texas as well as Oklahoma. In every new location, they make sure to hang a photo of Wade and his father with the family's first car. The photo is a testament to how far the company has come since it began as a used car lot on unwanted land back in 1956.

The father-son picture is also a reminder that each car purchase is the start of someone else's journey. And once the person heads out on the road, there's no telling what he or she can accomplish.

Deeper Insight

The "Our History" description here is 494 words — to me, it's the right length to tell the full company history and leave the reader satisfied. The company I describe here has "existed" since the 1950s yet I still managed to cover everything in 494 words.

That's because I didn't dwell too long on any of the three parts. I led with a memorable story, followed up with details on the company's growth and ended with a look to the future.

Even if your company was founded one or two years ago, you can still approach the "Our History" page the same way. Tell readers how everything started, explain how you built it up and let them know what's in store.

And if you have one to two photos of the company's beginning or growth phases, it's always nice to include them so people can view your story and success first-hand — and come to trust you more!

149

Services

Many businesses have a "Services" or perhaps a "What We Do" section of the website. Every company's "Services" section will have a different feel so it's impossible to provide a standard template everyone can apply to their businesses.

I think it would be more helpful to discuss best practices for the writing you do yourself. That way, you can explain your services the way you prefer but with several foundational rules to make the copy stand out.

Here they are:

1. Remember, less is more

No matter the topic, do your best to keep it brief. The goal on a "Services" page is to give the reader an overview of what you do rather than 1,000 words that dive too deeply.

Ask yourself: if I had two minutes to explain the particular service in a face-to-face conversation, what would I say? Anything past those two minutes might overdo it.

When you reach the end of the description, consider a link to a case study to keep someone's attention (template on page 154). That way, you take people deeper into your website and allow them to "reset" their attention span. Once the "case study" page loads, they're ready to keep reading and give you more of their time.

2. Quantify quantify quantify

Wherever you can add numbers, do so. Examples include number of clients served, money raised, money earned, percentages reflecting client satisfaction and years in business. Never be vague — attach numbers to strengthen your case.

Rather than write, "We have provided graphic design to several small and medium-sized businesses throughout the region," give people a number. For instance, "We have provided graphic design to over 150 small and medium-sized businesses throughout the region."

Without the number, you make the reader decide what "several" means. Does it mean 5? 10? 50? No, it's 150. Numbers help you impress and stay in control of your own message.

3. Testimonials can back up your claims

A short testimonial either above or below the "Services" description adds value to the page. On the flipside, a giant blocky paragraph for the testimonial will slow down the pace and potentially lose the reader's attention.

The right length is:

"Acme Corporation was the best choice for my new blinds. The team was fast, friendly and efficient — plus I saved hundreds of dollars!"

Toss in a strong quote, make it short and keep things moving. The purpose of the page is to impress in a narrow amount of time — not bog someone down with hundreds of words. And remember you can always provide longer reviews on a separate "Testimonials" page.

4. Use bold and larger font strategically

Bold and larger font are techniques that can keep the reader glued to your content. The key is to not overuse either one so the approach remains fresh.

It's **distracting if you bold too many words**. Then people **don't know** which words **matter the most.**

But if you use bold sparingly, it can provide a boost to your content and make words **pop.** See?

As for larger font, again it comes down to "less is more."

If you use larger font too much, it can feel like you're shouting at the reader.

Know what I mean? It's "loud" and distracting.

But if there's a line or phrase you want to emphasize, then set it apart with larger font. Ask yourself: does it feel like I have too many phrases in large font? If the answer is no, then go ahead.

Make it big and bold.

Q and As

On company blogs, businesses use Q & A posts to interview clients or staff members, share information about products/services or cover another topic altogether.

The key with every Q&A is to have a conversation rather than ask questions independent of each other. Why? The latter is boring. Here's an example of a Q&A for a new employee.

Uninteresting Q&A

1. **What's your name?**
   John Doe

2. **What's your role at Acme Industries?**
   Assistant regional manager overseeing franchise operations

3. **Where did you work before Acme?**
   I was a district supervisor at a sewage treatment plant in Arkansas.

4. **What are some of your hobbies?**
   Bike riding, reading mystery novels and playing with my two kids

————

See what I mean? The Q&A is flat and emotionless. That's because the questions have no common thread, and **it's not a conversation.**

Here's the same Q&A, but this time the questions follow a logical progression.

Interesting Q&A

1. **What's your name?**
   John Doe

2. **What's your role at Acme Industries?**

152

Assistant regional manager overseeing franchise operations

3. **You're new to the construction industry. What were you doing before Acme?**
   I was a district supervisor at a sewage treatment plant.

4. **Wow, what was it like working in a sewage treatment plant?**
   Let's just say it was a dirty job, but someone's gotta do it.

5. **True enough! What skills did you learn at your last job that you now bring to Acme?**
   So many different skills. I learned how to...

———

Do you see the difference? The second version is a **conversation** between two people. The first one feels rigid and impersonal. Plus, the second option is way more enjoyable to read. It has a comfortable, easy pace that flows along with little effort.

Remember: a Q&A is a dialogue between two people, and one question should spark the next.

As you construct a website Q&A, ask yourself:

▸ What piece of detail from the previous answer can I use to create the next question?
  ○ ex: "sewage treatment plant" leads to "what was it like working in a sewage treatment plant?"

▸ Do my questions sound like something a person would actually say? Or do they come across like corporate mumbo-jumbo?

Who are good people to feature in a Q&A? A few examples:

▸ New employee

▸ Long-time employee

▸ Devoted volunteer

▸ Happy client

Make sure each answer doesn't become too long; one to two paragraphs is enough. After that, the reader may lose interest.

Case studies

Case studies are the best place to demonstrate how you deliver for your clients. Too often, though, we forget to include details that make the study meaningful.

It's easy to tell the reader you do great work for clients, but the specifics of the project make you credible, believable and interesting. Also, remember you're telling a *story*, and it should keep the reader's attention all the way to the end.

The best way to lay out a case study is in six parts:

- Client testimonial

- Brief description of the project

- Staff involved

- The challenge

- The process

- The success

For the scenario below, imagine the company, ABC Events, handles event planning for corporate clients.

Client testimonial

Comes at the beginning to lure in the reader and give you instant credibility

> "ABC Events was a terrific partner for our annual company meeting.
> I recommend them for any corporate event, big or small!"
> — Derrick Johnson, CEO of Tech Corporation

Brief description of the project

Plan and execute a three-day company wide annual gathering for Tech Corporation, which manufactures car parts for retailers across the Midwest.

Staff involved

- Person 1, job title

- Person 2, job title

- Person 3, job title

- Person 4, job title

The challenge

In March 2016, Tech Corporation approached ABC Events with a unique situation: stage three straight days of catering and events for a company-wide annual meeting near Cleveland.

Tech Corporation had never done a three-day annual meeting, and its team was concerned an event planning company wouldn't be able to handle the logistics and all the moving parts (over 600 employees). We told them about the <u>many corporate clients we've worked with</u>, built Acme's trust and went to work. The annual meeting was only two months away, so we needed to act quickly.

Deeper Insight

- Introduce "The challenge" and include relevant people, places, dates, times and locations

- Lean hard on numbers (ex: three straight days, 600 employees) to underscore the scope of the challenge

- Link to past clients, if applicable, to flex your muscle and show who else trusts your services

- Set up "The process" with a final line that adds a little drama. So the reader thinks, "OK, what happened next?"
    - Ex: "we needed to act quickly"

The process

We began with a comprehensive planning session at Tech Corporation offices to understand the annual gathering — to be held at a summer camp near Cleveland — and set expectations for the role we would play. From there, we put our team in three clusters: catering, audio/visual and general logistics. Each group held weekly check-in calls with Tech Corporation so we covered every last detail.

The three-day event was a significant undertaking. We had to plan out 12 meals, assemble chairs and tables five different times (for 600 people) and set up audio/

visual equipment for seven different breakout sessions. A week ahead of the annual meeting, our team decided to meet at the camp site to scope everything out.

As the gathering approached, we were ready.

Deeper Insight

▸ Show examples of how your team operates (ex: comprehensive planning session, weekly check-in calls, visited the camp site a week ahead)

▸ Again, focus on statistics rather than be vague (ex: 12 meals, seven different breakout rooms)

▸ Include a line at the end to set up the final piece of the story ("we felt we were ready")

The success

The three-day annual meeting was nonstop action with people coming and going in all directions. But since we had prepared thoroughly over the previous two months, everything went off without a hitch. In fact, during one of the breakout sessions, in which we had planned for a single guest speaker, three people arrived to be on a panel. Because we plan for the unexpected, we had extra audio equipment and quickly created an audio system to support three microphones.

As for the catering? Each day, the hot food stayed hot. And the cold food (like late-night ice cream) didn't melt. Tech Corporation CEO Derrick Johnson and his team raved about our service and told us we did a fantastic job. In fact, they've already hired us for next year. Mission accomplished!

Deeper Insight

▸ Give specific examples of success (ex: one guest speaker turned into three, hot food stayed hot)

▸ Circle back on "The challenge" and explain how you found success despite the tough task (ex: did a fantastic job)

Staff bios

The staff bio follows a simple formula for everyone from recent grads to people with work experience.

1. The job title and role you currently have

2. Short description of the current job

3. Short description of past work/jobs

4. List of a few hobbies/interests (not necessary or always appropriate but helps to show your personality)

5. Your education (ex: where you attended college)

Your company might ask you to include other info or write the bio with a particular tone. Go with the flow but keep in mind the five-point list.

How to write a professional or executive bio — recent grad who's new to a job

(1) John Doe is a junior account manager at Acme Industries who supports various projects in our new acquisitions division. (2) He specializes in data analytics, social media management and event logistics. (3) In college, John co-founded AcmeGo+, a startup that monitored traffic flow across campus to help students avoid crowds and save time. (4) Outside of work, John plays classical guitar and grows vegetables in a small garden on the roof of his apartment. (5) He has a B.A. in philosophy from Big State University.

Deeper Insight

Let's go back to the five-point list.

1. The job title and role you currently have
   ► One clean line sums it up.

2. Short description of the current job
   ► Provide specific job duties (ex: social media management).

3. Short description of past work/jobs
   ▸ Again, specifics. Tell people not only *where you worked* but also what you did.

4. List a few hobbies/interests (not necessary but helps to show your personality)
   ▸ Instead of "guitar," I wrote "classical guitar." Instead of "likes to garden," I wrote "grows vegetables in a small garden on the roof of his apartment." Details make you more interesting!

5. Your education (ex: where you attended college)
   ▸ College comes at the end of your bio. It also should appear at the bottom of your resume. Lead with the job you have. That's the most important part of your bio and should come first.

How to write a professional or executive bio — someone with work experience

———

(1) Jane Doe is a senior account executive at Acme Industries who leads our renewable energy division. (2) She focuses on bringing solar power to new industries and oversees current solar projects. (3) Prior to Acme, Jane worked for Big Nonprofit on a team that improved the quality of drinking water in third-world countries. She also spent time at Little Nonprofit and studied the impact of oil spills on the seafood industry in the Gulf. (4) Outside of work, Jane likes to ride her bike around Houston with no particular destination and read mystery novels (she's also writing one of her own). (5) She has a B.A. in art history from Tech University.

Deeper Insight

Again, our five-point list.

1. The job title and role you currently have
   ▸ One clean line sums it up.

2. Short description of the current job
   ▸ Provide specific job duties (ex: oversees current solar projects).

3. Short description of past work/jobs
   ▶ Since Jane has worked for a bit, she described her role at the past two jobs.

4. List a few hobbies/interests (not necessary but helps to show your personality)
   ▶ Instead of "ride her bike," I wrote "ride her bike around Houston with no particular destination." Instead of "read books," I wrote "read mystery novels (she's also writing one of her own)." So much more colorful this way!

5. Your education (ex: where you attended college)
   ▶ Again, lead with the job you have. Finish out with education.

Photo captions

Photos on a website are more than visual elements; they can often be the only part of the page people read.

Why? A colorful photo can draw us in faster than a block of text. Then, our eyes naturally move to the bottom of the photo to read the caption. While it's nice to think people will read *everything we write,* we should assume in many cases the photo caption is our only chance to explain what the page is about.

What does an effective caption mean for your blog or website?

When you post a photo, make sure you tell people what's going on *and why it matters.* Then, the reader thinks, "Oh, *ok.* Now I know why I should care."

Other important pieces of info to include in the caption, when relevant:

▶ Specific location

▶ Day/month/year

▶ Person's first and last name

▶ Person's job title, if applicable

▶ Name of event and its explanation, if applicable

Here's a photo/caption from an event in which I spoke to a room of business professionals.

Danny Rubin, an author and speaker on business communication skills, speaks to marketing professionals at a workshop in Virginia Beach called "The LinkedIn Profile Crash Course." The event took place at the Rubin Communications Group offices on September 27, 2016. In the workshop, Rubin explained that the most effective approach to a profile summary is to write in the first person and explain why you're passionate about the work you do.

———

The caption contains all necessary information in case the person doesn't read the rest of the content on the page.

WHO: Author and speaker Danny Rubin

WHAT: "The LinkedIn Profile Crash Course"

WHEN: September 27, 2016

WHERE: Rubin Communications Group offices in Virginia Beach

WHY: Rubin explained that the most effective approach to a profile summary is to write in the first person and explain why you're passionate about the work you do.

Calls to action

An effective website takes visitors on an effortless journey from page to page. As much as possible, the content should flow like a conversation and have a natural progression.

Yes, there are plenty of other calls to action on a website. Phrases like:

- Donate now
- Click here
- Learn more

Each one can play a role, and I'm not saying do away with them completely. But when you can, look for opportunities to make the calls to action flow out of the content that precedes it.

Example 1

Often times, the "About Us" page is the first place visitors go after they land on the home page. They want a quick synopsis of the company/organization before they invest any more time.

A great place for a call to action, then, is at the end of the "About Us" description.

———

Final lines of "About Us" leading into call to action:

At Acme Corporation, environmental stewardship is at the core of everything we do. And we're proud to handle all of our production here in the US.

Then, the call to action:

<u>Learn more about our initiatives</u> to save water and energy at our US production facilities.

The line about "environmental stewardship" almost sets up further discussion on the topic, doesn't it?

It's an easy transition and opportunity to help visitors become more familiar with your operation.

Example 2

On the staff bios page, Acme Corporation has a bio for Ron Walters, a senior manager. Within Ron's bio (staff bios template on page 157), it reads:

Ron volunteers his time with Acme Community Organization, which provides meals for the homeless.

Then, at the end of his bio, start a new line and write:

<u>**Check out our blog post**</u> **on Ron's work with the homeless and how he inspired our entire team to be involved.**

Now, direct visitors to follow you over to the company blog and learn more about your community impact. The call to action isn't pushy or forced; it makes sense in the flow of the content.

Example 3

Calls to action don't need to happen at the end of the section of content. They can appear in the middle of a content block too.

Let's say on the Acme Corporation page about environmental stewardship, there's a section about how the company conserves water.

_____

At each production facility, we are able to collect and use rainwater, which allows us to rely less on the city's aquifers for our supply.

**Learn more about our green-certified campus** *and how it's become a model for other companies here in Montana.*

Our production process has four steps. The first step is…

_____

If there's additional information visitors should know, place an italicized call to action in the flow of the content on the page or use an image to break up the text.

It's one more way to take people deeper into your site and build their trust.

Meta descriptions

Meta descriptions (information below a website page on search engines that summarizes the page's content) are a moving target when it comes to the perfect strategy.

It depends on how much weight Google and other search engines place on the description when it delivers search results.

My recommendation is to search the latest on meta descriptions at the time you read this section and see how to write them best. It's critical to include meta descriptions because they encourage people to click through to your site.

So many aspects of classic business writing do not bend with the times. Meta descriptions, however, are not one of them.

How to write blog posts

SEO FOR COMPLETE DUMMIES (LIKE ME)

I'm a writer, not a computer whiz or search engine optimization (SEO) expert. That's why in 2012, when I first started to blog and explore the mystery of search engines, I was fortunate to sit through a presentation in which author and speaker Jay Baer made SEO nice and simple.

Jay said, "For your website, SEO means you should provide useful content and solve someone's problems."

That one piece of instruction allowed me, over a period of three years, to hone my message, build website traffic, develop a niche, write books and establish a brand.

Glad I was listening that day. :)

I learned search engines want to find the most honest, authentic, no BS answers out there. Commit to being that reliable resource and, over time (may take several months), you will see the traffic come your way.

What's your business? What do you specialize in?

OK, now think to yourself: what are the questions I hear most frequently from my customers? What don't they understand? Where do they struggle?

Take five minutes and write down a list of 8–10 questions — the **most** critical ones you can answer.

Go ahead and do it now. I'll wait.

...

...

...

All right. Got your list? Cool. Each one of those questions should become a blog post on your website. Let's say you run a vinyl record shop. Here's a question you might hear:

Is it OK to leave vinyl records stacked on top of each other?
Perhaps that's a question a lot of people ask you. You can even call the blog post the exact question in bold. And then put the key terms in the url too.

Is It OK to Leave Records Stacked on Top of Each Other?
(www.acmevinylrecords.com/leave-vinyl-records-on-top-of-each-other)

Then, you need to provide a great answer. I mean, a *really great one*. The best the internet has ever seen. If you're the expert, you can do it, right?

Here's the beauty of SEO for non-tech people. If you provide the best answer, people will share it with others. As your views increase, the content will move up the organic rankings (not referring to paid search results). You provided the greatest solution for a given query.

Answer the question to the best of your ability. The keywords (ex: "vinyl records stacked on each other") will likely appear within the post in a natural, conversational way (what search engines prefer, anyway). Yes, include the key terms in alt tags and meta descriptions too. It's a moving target how much those sections matter, but you should always have keywords there.

But mostly, speak from the heart, share great advice and answer people's questions as well as you can.

That's SEO for dummies. How'd I do?

To help you along, I provide even more blog ideas and templates on the following pages. Now you can't say you have writer's block — my suggestions will keep you busy for a long time!

10 different blog topics to help you generate ideas

Remember: with every post, you must provide advice that solves someone's problem. In doing so, you prove you're the expert on a given topic.

- Answers to your five most frequently asked questions (or one blog post for each of the five)

- Lessons learned while working on a client project

- Step-by-step instructions for a particular task (something you find yourself explaining to people all the time — put it in writing!)

- Updates on industry rules and regulations (be the company "in the know")

- Recaps of your experience at big events/conferences

- Recaps of presentations you gave either internally or to external audiences

- Video tutorials on popular topics

- Photo gallery from an event/project

- Common misconceptions about aspects of your business ("You think 'X' should be done this way, but actually you need to do it like this")

- Tools/resources you recommend others use too

Advanced lesson: Thinking about writing a book on a topic you know well? Here's the way to do it with maximum efficiency.

As a first step, lay out your table of contents (TOC). For reference, look at mine in this book or in *Wait, How Do I Write This Email?* Then, go through the TOC one by one and write blog posts on each topic.

Over time, you will fill out your blog with smart, insightful content (that also doubles as great material for SEO purposes). At the same time, you will assemble your book — one blog post at a time.

Gather all your blog posts, match them to your TOC and there you have it: a book manuscript.

The secret to effective blog post headlines

I first came to understand online headline writing by accident.

At the time I wrote this section of the book, the search terms "apply for a job company isn't hiring" shows my blog post in the #1 position (guest post for Business Insider).

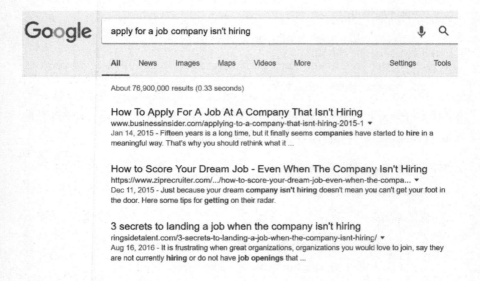

Because of the prime position on Google, the post has been viewed thousands of times. The original post on my blog also receives steady organic traffic.

A **weaker headline** would have limited the success of the blog post. Here are the four steps I followed to maximize impact.

1. Answer the question

Your blog posts, as much as possible, should help people solve problems. You want to offer how-to guides, insights and instruction. Obviously, people google answers for all kinds of topics. Whatever you know best, provide the answers on your website.

In my case, I dispense practical business writing instruction. That's why I wrote an email template to apply for a job at a company with no open positions. I thought to myself, "How would someone search this topic on Google?"

I didn't try to be cute and clever with something like, "No Job Posting? No Problem!" Odds are, few people search those phrases on Google.

Instead, I went with a more likely string of words: *How to Apply for a Job at a Company That Isn't Hiring*

2. Put key terms in the URL

Your content management system (ex: Wordpress) may automatically place the words from your headline in the URL. Business Insider used /applying-to-a-company-that-isnt-hiring. They kept the main words of the post in the URL, and it helps for search purposes.

3. The right amount of words

Google will generally display 50–60 *characters* (not words) of your headline. To make sure your headline fits completely, use an emulator tool to count your characters.

Google shows my entire headline in the result — nothing is cut off. Again, I don't know if I would be "penalized" for a longer headline, but I do know the one I have is easy to read and understand. The human element matters here too.

4. Don't play the "stuffing" game

In the column itself, I don't repeat the phrase "apply for a job at a company that isn't hiring" like 20 different times. Google hates that. It's inauthentic and a way to "game the system." Instead, share your answer in a natural, organic way like a conversation to a friend. The key terms will likely be part of your answer, anyway.

———

To recap, a winning blog post headline:

1. Answers a question the way a person would search it on Google

2. Includes key terms or phrases from the headline in the URL (and meta description)

3. Uses standard rules of capitalization

4. Stays within 50–60 characters so the headline is entirely visible

5. Doesn't "stuff" the post with the exact headline text over and over; instead, it's conversational

Opinion-style blog post

One of my favorite sections of a newspaper is the opinion pages. Writing opinion pieces is hard, and I respect columnists who make a strong argument and keep my attention from start to finish.

While every columnist has his/her own style, many follow a formula we can apply on our own blogs and websites. In a nutshell, start with a story, state your argument, defend your argument, offer a solution and conclude with a reference to the story from the beginning.

As an example, I use excerpts from the blog post I discussed in the last section ("How to Apply for a Job at a Company That Isn't Hiring").

Opening section: Start with a story or situation

Open with a quick story or situation that frames the argument. What spurred you to have the opinion? Was it an event you attended, article you read or something you saw on TV? The story grabs the reader and puts your opinion in context.

Opening section example:

Welcome back, economy.

<u>USA TODAY reports</u> *employers added an average of 246,000 jobs each month in 2014, the best year for job growth since 1999.*

Fifteen years is a long time, but it finally seems companies have started to hire in a meaningful way.

Second section: Make your case

With the story/situation as the backdrop, lay out your argument.

Second section example:

Spurred on by the positive job market, you should do the unusual: apply even when there are no open positions.

Third section: Defend your case

Tell the reader why your argument holds water.

Third section example:

First of all, what do you have to lose? Nothing. Exactly.

Second, what if your email pitch is compelling and puts you on the company's radar? Then if the boss does need to make a hire, you come to mind.

Fourth section: Offer a solution

Give the reader an action item.

Fourth section example:

I created a template to "apply" for a job even if the company isn't hiring. You never know where a single email can lead.

[And then I included the entire email template AKA the "action item."]

Fifth section: Make your closing argument

Reiterate why your reasoning is sound.

Fifth section example:

You can wait for opportunities to find you (you'll be waiting a while) or you can go out and grab them. A polished email introduction to a company could lead to an interview and change the entire course of your career.

Sixth section: Return to your story

Circle back on the story from the beginning to "bookend" your column — the same theme at the start and finish.

Sixth section example:

As I said, the job market is hot. No better time to take a chance than right now.

Deeper Insight

In the example above, I offer step-by-step writing instruction. Certainly your opinion blog post doesn't need to provide such detailed advice. But it's still important to frame the argument with something that happened in the world, make/defend your argument and then offer a solution(s).

Profile on a volunteer

"Profile" stories are a smart way to bring your website content to life and spotlight the people who power your company or organization. They also become ideal vehicles to promote your services and offer "calls to action."

The goal is to grab readers right away and allow them to "meet" the person who is the subject of the post.

No matter your product or service, you can share profiles of your staff, clients, investors, volunteers and anyone else integral to your success.

Here's a fictitious "profile" blog post about a volunteer, step by step. The person in the blog post, Alicia Hammonds, is a volunteer for Big Nonprofit Association.

Blog Title: Getting to Know [First and Last Name], [Job Title/Relationship to the Company] at/with [Name of Company/Organization]

Opening section: Place the person in a situation

Open with a quick story or situation that puts the person in his/her element. How would you sum up the person's role and overall impact? The story grabs the reader better than a line like, "Today we are profiling ____…" Booriiiiing.

Opening section example:

If it's Tuesday morning, you can bet Alicia Hammonds is in the food pantry sorting cereals, cans of vegetables and bags of rice. For the past four years, Hammonds has been one of the most active volunteers at Big Nonprofit Association, and we rely on her each Tuesday to keep the pantry organized and prep food for the 14 homeless shelters we serve.

Second section: Provide context for the person's hard work

Step back and put the person in context with the rest of the company/organization. Be specific with clients you serve and provide numbers when possible to help the reader understand how much, how many and how often.

Second section example:

Hammonds is one of 50+ volunteers we depend on so Big Nonprofit Association can reach more than 2,000 people in need throughout the Denver community. Our food closet benefits organizations like Nonprofit A, Nonprofit B and Nonprofit C. We like to think we're the engine that makes everything go, and our volunteers play a huge role in that task.

Third section: Share a specific example about the person

By now, you have introduced the person and given context on the organization. Now share what makes him/her exceptional.

Third section example:

Back on February 22, 2015, Hammonds was called on for a critical task during a powerful blizzard: catalog all the food in a narrow, four-hour time window before it was rushed out the door to aid homeless shelters. With focus and determination, Hammonds completed the task and our community partners had the supplies they needed to care for those left out in the bitter cold.

Fourth section: Put the specific example in context

Explain how the short story about the "profile" person exemplifies who you are as a company/organization.

Fourth section example:

Hammonds' quick work on that frigid day in February typifies the effort we receive from all of our volunteers. Everyone understands their roles at Big Nonprofit Association and takes their jobs seriously. We're lucky to have so many selfless people who make a difference in Denver every single day.

Fifth section: Call to action

Use the final section to encourage the reader to contact you for new business or another kind of inquiry.

Fifth section example:

Would you like to become a volunteer for Big Nonprofit Association? Fill out this form and please allow us 2–3 days to respond. Thanks for your interest!

————

Profile on a staff member

Here's a fictitious "profile" blog post about a staff member. The person this time, Jovan Mitchell, is the assistant general manager for Acme Corporation, which owns several food trucks that sell to patrons across Philadelphia.

Blog Title: Getting to Know [First and Last Name], [Job Title/Relationship to the Company] at/with [Name of Company/Organization]

Opening section: Place the person in a situation

Lunchtime on a Wednesday in University City. A line 50 people long gathers around the Acme Taco food truck as employees inside the truck work feverishly to meet demand. Out of the back of the truck emerges Acme Assistant General Manager Jovan Mitchell. He's checked on Acme Taco and is now headed across town to monitor a different food truck. Another day in the life of "the truck boss," as our team calls him.

Above, I open with a story, grab the reader's attention to show what "a day in the life" is like for a staff member.

Second section: Provide context for the "profile" person's hard work

As assistant general manager, Mitchell is responsible for our 11 food trucks across the city that do business with over 500 people each day between the hours of 11 a.m. and 2 p.m.

Back in 2011, Acme Corporation had one food truck and two employees. Today, with 47 full or part-time employees, we're one of the fastest growing small businesses in Philadelphia. And we rely on Jovan every day to keep watch on it all.

Note the details in the story: 11 food trucks, 500 people, 11 a.m. to 2 p.m., 47 employees. The numbers make Jovan's work even more impressive.

Third section: Share a specific example about the "profile" person
Back on April 9, 2015, Jovan's talents were on full display. We had five employees out sick, one truck with a flat tire and another that overheated. Mitchell reassigned the team to staff every truck and coordinated roadside assistance for the vehicle repairs.

Fourth section: Put the specific example in context
In the background that sunny day in April, it was crazy. To the customer, it was business as usual. And that's why we love Jovan. He's never rattled or off his game. We know when problems arise, he will come through.

Fifth section: Call to action
Like our blog post? Subscribe to our newsletter and never miss an update — plus, weekly discounts and coupons!

Deeper Insight
Don't forget to include at least one photo of the staff/team member on the blog post page!

Post that plays off of a story in the news
I'm a huge fan of a company blog because it's a space to show your team is living, breathing and out there doing cool work. *Without a blog, the site would likely be static with nothing new to share or report. In other words, it looks like no one's home!*

That's why, alongside opinion columns and profiles of team members, you should relate the work you do to the day's headlines. It shows you're current, "with it" and thinking about how your services apply to real-world issues.

Here's a blog post example for Acme Daycare, which runs a chain of daycare centers in Seattle. A few days ago (fictitiously), the national news reported on a flu outbreak at 47 daycare centers across Texas and Louisiana. Now, Acme's corporate team wants to post a blog to ensure parents its facilities are clean and germ free.

Blog Title: Happy, Healthy and Germ Free: Why You Shouldn't Worry About the Recent Flu Outbreak

Opening section: Reference the news item right away to put the blog post in context

By now, many of you have <u>seen the news</u> about the dozens of daycare centers in Texas and Louisiana with a flu outbreak. Health officials believe the flu began at a daycare in Dallas but has since spread and impacted hundreds of children.

Note: Link to the news item so the viewer can read further.

Second section: Explain why the news item is relevant to your business

We want parents to know we have not seen any traces of the flu in our facilities and are working to make sure that remains the case. We know your child's health is priority #1 — for you and for us. And while we always maintain the strictest standards of cleanliness in our centers, we're now being extra cautious given the outbreak in Texas and Louisiana.

Third section: Provide the reader with tips as they pertain to the topic in the news

To help us keep everyone safe at Acme Daycare, please follow these health tips before and after your child attends one of our centers:

- Tip 1
- Tip 2
- Tip 3
- Etc...

Fourth section: Call to action

If you have any questions or believe your child may exhibit flu-like symptoms, please call your Acme Daycare location and detail the situation with the on-site nurse. He/she will determine if your child should come in or stay home.

Thanks for your cooperation.

– The Acme Daycare Team

Breakdown

I chose a news example that involves sickness so the topic has a bit of an "emergency" feel. Keep in mind you could relate all kinds of news stories back to your business that aren't time sensitive or don't threaten someone's health. Examples:

- ▸ Viral video that has everyone talking

- ▸ Court decision, government act or recent election

- ▸ Local community initiative or program

- ▸ Impassioned opinion column in the newspaper that has people riled up

If you keep close watch on the news, you will find topics that impact or relate to your business in various ways.

Again, blog posts with news relevance make your website feel "alive" and like a place worth visiting. And that's a great reputation to have in the minds of consumers.

Post on a recent success

Finally, here's a template for a blog post about a company/organization success story.

You should always use your blog to share the good work you do. It's great content to share with your audience and also becomes excellent reference material when you want to prove your worth to future business prospects. ("Hey, let me send you a quick link to a recent success we had.")

Blog Title: Acme Nonprofit Gathers Over 200 Volunteers to Clean the Smith River

Opening section: Reveal the recent success right away so you don't make the reader search around

The team at Acme Nonprofit is proud to have led a major volunteer effort on Saturday, June 9 to clean a portion of the Smith River.

We gathered over 200 people from across the region to aid in the clean-up. From children to seniors, everyone put on rubber gloves and did their best to remove trash and other debris from a section of the river long ignored by our community.

Check out our photo gallery!

Second section: Explain why the recent success matters (why should readers care?)

For the past decade, the Smith River has been an eyesore and a place many in our area tried to avoid. Plus, the pollution has depleted the local fish population.

That's why Saturday's cleanup was critical to not only the local wildlife but also the broader community. Everyone benefits with a healthier, cleaner environment.

Third section: What are the next steps? Where does the "success" go from here?

While we made a lot of progress on Saturday, there is still work to do. We have plans for additional river cleanup days in July and August and will release those dates soon.

We are also evaluating other polluted locations in the community that need our attention.

Fourth section: Call to action (How can you engage people further?)

To stay in the loop on upcoming projects, please join our email list!

Breakdown

The blog post here isn't long. You may certainly have additional content to add. But for the purposes of laying out a template, the four sections get the job done.

We open with the "main point," explain why the project matters, provide any next steps so people know what's to come and then finish out with a clear call to action.

No matter the "recent success," this model will help you write the information in a way that's clear and follows a logical progression.

Landing pages

In the "Extra stuff" section, I will not provide step-by-step templates because there are countless directions you could take depending on the topic.

The copy depends on your product/service and what you want from the reader. Are you selling an online course? In-person class? Consulting? Coaching?

Landing pages come in all stripes. Some are brief with a simple sentence or two and compelling image. And in many cases, that approach gets the job done.

But if you're selling an online course or attempting to gain new leads, your "pitch" should be more robust and detailed. In that case, follow these rules to ensure the reader stays with you from top to bottom.

Include your headshot: Let people see your smiling face. It's a guaranteed way to build trust and authenticity. And make sure it's a professional photo — NOT a casual pic from Facebook!

Tell a story of success: Show potential customers/clients you have already delivered for someone else. The story should contain a beginning (challenge/obstacle), middle (how you overcame the challenge/obstacle) and end (the positive outcome).

Early on with my free webinar called "How to Write the Best Job Application of Your Life," I used a story on the registration page (which served as the landing page too).

In the story, I discussed how I helped a young woman stand out in the job market. I include the anecdote below.

To register for the free webinar yourself, visit DannyHRubin.com/jobapp webinar.

———

To prove I'm worth your time, let me tell you a quick story of a person I helped to land a job — a temp employee at a chicken plant.

[Beginning — challenge obstacle]
Recently, a young woman named Jasmine reached out for career coaching help. She worked at a chicken plant in Maryland as a temp employee in the HR division. She wanted a full-time job BADLY but couldn't get noticed in the job market. I could tell she was frustrated and in a tough spot.

[Middle — how she overcame the obstacle]

I asked Jasmine about a time in her temp job where she demonstrated work ethic and hustle. She then told me a story...a GREAT story...and we used the experience in her resume and as a central theme in her cover letter.

[End — the positive outcome]

The result? THREE job interviews and TWO subsequent job offers within a MONTH. She then accepted a position in HR at a Washington, DC, law firm. What a terrific opportunity!

Additional landing page strategies:

- ▸ **Provide info on your background or the team's background:** Share your bio and credentials so people feel comfortable with who you are and what you offer. For help, refer to my "Staff bios" section on page 157.

- ▸ **More is more:** With a landing page, don't be too brief with the content on the page. Go in-depth on your course, consulting services, etc... and anticipate the viewer's questions/concerns. Never let the person leave the landing page because they were confused or under-informed.

- ▸ **Testimonials:** Gotta have 'em, and it's even better when you can show the person's face too. A good rule of thumb is to have at least three smiling people who can sing your praises. And keep the review quote to one to two sentences max. If the testimonial goes on too long, you could lose the reader.

E-newsletters

What are the best writing strategies for e-newsletters?

With each e-newsletter, we need to respect the reader's time and encourage clicks as soon as possible. After all, the e-newsletter is a portal into your website and a way to give the person a deeper look at the work you do.

Every e-newsletter is different so please take my advice as a general outline to be woven into your current setup.

Here are my best practices for e-newsletter communication:

- **Subject line is EVERYTHING:** Please, please, please do not call your newsletter something generic like, "Acme Corporation May Newsletter." It will doom the success of your newsletter. Here's why:
  - The subject line isn't engaging.
  - The subject line keeps people from opening the newsletter and seeing your great content.
  - Instead, think how you can "tease" the content inside the newsletter with a compelling subject line. Ask the reader a question or use a pronoun to create a sense of curiosity ("We have never seen THAT before!"). And then the reader thinks, "Never seen what???".

- **Less is more:** Don't fill the newsletter with too much text. Tease the content and provide clear "call to action" links so the person can continue reading and gain additional information.

- **Be conversational:** Remember an e-newsletter is a one-on-one conversation with the recipient. It's not a formal exchange between you and a big group of constituents. Let your guard down and show personality — the approach will give off the right tone.

- **Don't overdo it:** If you send out a monthly e-newsletter, for example, don't stuff the email with 15 different items. It's too much and will overwhelm the reader. On the whole, five to six pieces of content is a happy medium. It's enough to give people options but not too many that they become overloaded.

- **Tread lightly with images:** In my experience, too many images in an email can hurt your deliverability. Email services could view the message as spam never to be heard from again. My rule of thumb is to use images for effect and a way to break up a long stream of text. A text-first approach is your best bet to make sure the email reaches the intended inbox.

Ask for feedback

To perfect your e-newsletter, poll your audience on what it wants.

In 2016, I sent a survey to my own audience (at the time, it was 7,000 subscribers) and asked what people liked/didn't like. My e-newsletter is called THE TEMPLATE

(DannyHRubin.com/thetemplate), and each week I offer practical, step-by-step communication tips.

I learned so much from the reader survey, especially the aspects of my e-newsletter people appreciated the most.

One of the questions: **Which part of THE TEMPLATE do you like most?**

- Danny's intro message [9%]

- News to impress your co-workers [12.8%]

- Smart productivity tips [50%]

- Free writing templates [23.1%]

- Danny's video clips [3.8%]

- What Danny is up to in the week ahead [1.3%]

I was surprised to learn the #1 response — and it wasn't even close. Fifty percent of respondents told me they like smart productivity tips (links to top productivity articles and columns from the past week).

OK, great. Prior to the survey, I included one to two productivity links. Armed with the survey knowledge, I beefed it up to four to five links, and I made the section more prominent within the e-newsletter.

A month later, I received this note from a reader:

> Hi Danny,
>
> Just thought I'd send some feedback on your TEMPLATE newsletter, since you've been using the new format for a few weeks. I like it! :-)...Overall, your new format has made THE TEMPLATE more useful & efficient.

When in doubt, ASK. The marketplace will always tell you what it wants.

My Journey: Part 6
That time I contacted the same organization 30 times

There's a saying in advertising that consumers need to see your message seven times before they act on it.

When it comes to promoting a book, make it closer to thirty.

I called a lot people at schools and organizations about *Wait, How Do I Write This Email?* I always made sure to introduce myself properly (page 211) and plan out my talking points.

I had a strong pitch, and people were generally receptive.

Then I would send a couple of books in the mail and plan to follow up a couple of weeks later to gain their feedback.

And that's when the "fun" began.

Email check-ins. Trying the person's office line to catch him/her at the office. More email check-ins. More phone calls.

Not all in the same day, mind you. Over a period of several days or weeks.

And that's how, with at least one organization, the follow-up efforts reached thirty times — until, at long last, the organization decided to make my book a priority and move forward with an agreement.

Nothing comes easy as an entrepreneur. But I believe in my book and its purpose — to help people write better so they can open doors in their careers.

That's why I wasn't dejected each time someone brushed me off or ignored my follow-up requests

No, I was emboldened. Inspired, even.

I thought, "If I only check back one more time, *then* it will happen."

What did I learn through the process? That persistence pays off.

What else did I uncover? That I wasn't even close to "figuring it all out."

The journey continued, and the next step was a big trip to The Windy City.

Chapter 7
Fundraising Tactics

INTRODUCTION

In the world of fundraising and development, writing is critical.

Your words should capture the reader's emotion and compel the person to donate or volunteer. That's why I believe effective writing for fundraising hinges on two key ideas:

- Storytelling
- Authenticity

In the following pages, I show you how to weave both concepts into common writing challenges for fundraisers — both online and offline.

Grantwriting

WHY YOUR APPLICATION NEEDS A POWERFUL STORY

Everyday, thousands of organizations ask for money either from the public or charitable entities. With so many groups in need of funding, what sets your "pitch" apart from the rest?

The answer: a powerful story.

In this section, I will focus on storytelling as it relates to a grant application. Yes, the funding organization must read about the purpose of a project/initiative and estimated costs — the basic facts that make your application worthy of consideration.

But the funder also needs to understand the heart and soul of the project; in other words, how your organization has made a difference in someone's life or *will* have an impact once the program begins.

Without the story, it's a bunch of facts and figures. *With* a story, you put a face to the mission.

There are several places in an application where a short story makes sense and adds value. Look for sections called:

▸ Statement of Need

▸ Project Description

▸ Participants

▸ Previous Outcomes

In the template below (fictitious), the organization, Acme Nonprofit, provides educational opportunities for children of immigrants through an initiative called Our Children, Our Community. The program is in its fourth year and has requested $15,000 from the Acme Foundation to continue the effort and increase the number of children it reaches.

PLEASE NOTE: Your story must align with the mission of the funder. In the anecdote below, Acme Nonprofit describes the experience of a young immigrant because the funder, Acme Foundation, wants to support organizations that help immigrants adjust to life in America.

Keep an eye on the footnotes as I explain each section at the end.

"A Brand New Start: How Our Children, Our Community Changed the Life of Maria Lopez"

(1) It's October 2016, and Maria Lopez sat in timeout for the sixth time in only the seventh week of school. Frustrated and rebellious, teachers couldn't seem to the reach the fourth grader and improve her behavior.

(2) Maria is a first-generation American. Her parents are from El Salvador and speak little English. That means Maria comes to school every day with limited English skills. What's worse, Acme Elementary School (where she attends) lost its funding for additional programming for ESL (English as a Second Language). Without a tutor, Marias regular teacher must lead the class and provide ESL lessons for Maria. In short, the teacher is overworked.

(3) That's where Our Children, Our Community stepped in. With the money we raised in spring 2016, we were able to secure a tutor for Maria three days a week (Mondays, Wednesdays and Fridays from 2–3 30 p.m.). Within a few weeks, Maria's teacher and guidance counselor could see a major improvement in her behavior and class participation.

(4) As Maria gained confidence with her English skills, she smiled more, paid attention and followed class rules. Her teacher even remarked that one time Maria assisted a classmate (and native English speaker) with a lesson on verb tenses. The teacher told us, "I had tears in my eyes watching Maria help someone *else* with his English. She's come a long way!"

(5) In October 2016 (during the period of multiple time-outs), Maria was close to failing in every subject. By mid spring, she had a B– or B+ in each class. We at Acme Nonprofit think that's a huge accomplishment.

(6) Maria is the reason the team at Our Children, Our Community works so hard. We know every child has the potential to be great. The key is to surround the child with a support system that addresses the need head-on and gets results.

(7) We are excited to continue our initiative for a fourth year and hope we can count on the Acme Foundation for the $15,000 grant. In 2017, we have a goal to help between 50–60 children. By comparison, our program assisted 26 children in 2016. We know our costs will increase for staff, supplies and other resources.

(8) When you consider our application, think about Maria and how she's now flourishing and excited for fifth grade. Help us make a similar impact on even more children.

———

Again, the story above is *part of the grant application* — not the entire application. But you can see how the story provides depth and allows the readers to understand how the grant will make a difference.

Here's the breakdown following the numbers associated with the story.

1. Place the reader into a situation immediately. Set the scene and explain the problem ("...teachers couldn't seem to the reach the fourth grader and improve her behavior").

2. Provide details on the person's life and nature of the situation (Maria has two parents who don't speak English, her school lost ESL funding and her regular teacher doesn't have time to create special lessons). The details in section two are crucial; the more specific you are about the person's situation (and how tough it is), the more compelling your application.

3. Explain the impact your organization made and be as specific as you can. Note how I wrote "...we were able to secure a tutor for Maria three days a week (Monday, Wednesday and Friday from 2-3:30 p.m.)"

4. Provide a detailed example of your efforts (in essence, a story within a story).

5. Include metrics of how the situation has improved (data points will be different given the situation; in the story here, the metrics are Maria's grades).

6. Explain how the person you profiled (ex: Maria Lopez) is a perfect example of why you care about the work so much.

7. Discuss the reason for the grant request and why you need the amount requested ("In 2017, we hope to help between 50–60 children. By comparison, our program assisted 26 children in 2016.")

8. Conclude with a reference once more to your story.

A few more points:

▸ Even if the application doesn't ask for a story or anecdote, you should still provide one. People love to read stories and your example will set the grant request apart from the others.

▸ Many online grant applications have character limits so you may need to exercise brevity as you tell the story. Always ask yourself, "Is this word, phrase or sentence necessary?" If the answer is no, cut it out.

▸ Make sure you don't include sensitive or private information about your story subject(s). When in doubt, LEAVE IT OUT.

Stories are great annual reports

You know that dynamite story you wrote for the grant application?

Copy and paste the story and drop it right into your organization's annual report. Why? **Stories bring your annual report to life.**

Yes, the data matters and you need to remind donors about the nuts and bolts of your work. But success stories underscore the kind of impact your team makes day after day.

The annual report should resemble a magazine full of powerful anecdotes and stirring photos. That's how you tug on the heartstrings of those who give (or could give).

Annual report = Book of unforgettable stories

How to rely on strong detail to make your case

If you ask someone for money, you better make a darn good argument as to why you deserve it.

In a grant application, the fine details allow the "darn good" argument to take shape.

Wrong way

At Acme Community Group, we have helped many children the past several years improve their grades and attend college.

Can you spot the vague, inexact language? I can.

▸ "many children"

▸ "past several years"

▸ "improve their grades"

▸ "attend college"

Yep, pretty much the entire sentence is a loss. While it's nice Acme Community Group helps young people succeed in school, the reader has no idea HOW successful it is. Why would the charitable organization provide a grant without clear evidence of what Acme will do with the funds?

Right way

At Acme Community Group, we have helped 147 children succeed in the classroom since our organization began in 2005. Of the 147 children, 85 percent improved at least one grade letter in every class where we provided tutoring. As well, 94 percent of the children graduated on time and were accepted into four-year colleges.

See the difference?

▸ **147** children

▸ **85** percent improved at least **one letter grade**

▸ **94** percent graduated on time and were accepted into **four-year colleges**

If you do great work in your community, then you already have the stats to back it up. Now, **use the data to make the case you're worth the grant money.**

As you draft the grant application, ask yourself:

▸ How many people do we help?

▸ What are the metrics that define success?

▸ What's the clearest way to provide the metrics?

▸ Am I backing up every claim of success with my metrics?

Even if it requires extra effort to track down your metrics, make it happen. You're asking for someone else's money to further your mission. Work like mad to quantify the job you do and leave nothing to doubt. **Nothing.**

Email templates

HOW TO ASK IF A COMPANY WANTS TO BE A SPONSOR

If you want to email an individual business owner about sponsoring an event, make sure the message is customized and genuine. Here's how.

Subject line: Sponsorship of [name of event; for instance, "Chamber of Commerce awards luncheon"]

Hi [person's first name] or Mr./Ms. [last name],

> Note: To determine if you should use first name or Mr./Ms., refer to my chart on page 16.

I'm [first and last name], a/an [job title] with [name of company/organization]. I hope you're doing well.

[Next, include a specific reference to the company's website. Look at recent projects and commend the person on a job well done. For instance, "Congrats on the completion of the multi-purpose center at Acme University. What a beautiful building!"]

I'm writing to invite [name of company] to be a sponsor of our [name of event] on [day and month; for instance, "Wednesday, May 21"] at the [location]. As a sponsor, we can offer your team [what the company receives; for instance, "the chance to display your logo in the event booklet and on the large projection screen above the stage"].

Note: See above how I make the "ask" high up in the email. "I'm writing to invite..." is the first line so the reader doesn't need to scan the email and think, "What am I being asked?"

We expect over [how many?] business and community leaders to attend the lunch. The event is an excellent opportunity to expose your business and brand to people from across the region [or if it's more of a niche event, you can write something like, "...to people within the local startup community"].

I have attached the sponsorship information sheet. Please let me know if you have any questions.

Thanks for your consideration,

– Your first name

Email signature

Deeper Insight

Lead the email with a reference to the company's website. That way, your email doesn't feel spammy and as though you sent the same message to 500 people.

I do understand there are times when you'll need to send sponsorship-related emails to a huge group at once. On those occasions, you can't customize the message.

But if there's a particular company you want as a sponsor, take the extra time and draft an authentic message. It will mean more, and you'll have a greater chance at a positive response.

How to send fundraising outreach emails

For the template here, let's imagine Jane Doe works for Acme Nonprofit and needs to reach out to past donors or people who have attended recent events. The goal: to make each person consider a gift of some amount.

Actually, the *real goal* at the onset is for the person to respond. Once Jane opens the conversation, then she may be able to win him/her over.

So what's the best way to encourage a response? *In short, the details make the difference.*

Here's what the email can look like. Also remember the same tactics apply for a physical letter too.

Subject line: We have the momentum. Now, we need you too.

Hi [person's first name],

I'm Jane Doe, a development associate for <u>Acme Nonprofit</u> here in [city]. Nice to meet you!

First, I want to say I'm impressed by [example from the person's career; for instance, "what you've been able to build at Acme Corporation. It looks like you provide IT services to some major companies, and I'm sure it must be fun to work with Company XYZ. Congrats on your success!"]

I'm writing to tell you what Acme Nonprofit accomplished in 2016 and what's on tap for the year ahead. In short:

- Accomplishment 1 [quantify success and link to examples as much as possible; for instance, "<u>we built 22 more playgrounds</u> in underserved communities, which will impact the lives of 1,300 children"]

- Accomplishment 2

- Accomplishment 3

And we have even bigger plans for 2017, which include [1–2 examples of what's to come or other ways to excite a potential donor].

Please let me know if you have a few minutes to learn more. I'm happy to find a time that's convenient for you.

Thanks,

– Your first name

Email signature

Deeper Insight

Notice how Jane opens the email. Right away, it's customized.

First, I want to say I'm impressed by what you've been able to build at Acme Corporation. It looks like you provide IT services to some major companies, and I'm sure it must be fun to work with Company XYZ. Congrats on your success!"

How did Jane "find" the (fictitious) information on the person's career? She googled the person's name, learned about his company and used what she discovered to enhance her fundraising email. Now, the email is tougher to ignore because it's not the exact same message to every person on her list. It's personalized and probably surprised the recipient because he didn't expect it to contain any info about his own life/career.

Then, Jane puts Acme's three best achievements in a bulleted list for better readability. She leans hard on numbers ("22 playgrounds" and "1,300 children") to prove success.

Finally, Jane mentions one to two goals/plans for the coming year. Here is another spot in the email you can customize. Let's say, in her research on the person's career, Jane uncovered how the IT company has a robust internship program and likes to teach young people important tech skills. It so happens Acme Nonprofit also strives to connect young people with technology-based careers.

Then, the line in the email could be something like, "And we have even bigger plans for 2017, which include a new initiative to teach underprivileged children how to code and build websites. I see Acme Corporation has a big internship program so you might find our program interesting."

True, you may not always find a direct connection, **but until you do your research, you never know.**

Does my email template take longer than a standard email blasted out to 100 people? Of course it does. But most of the template is the same for every person; you only switch out the info on the person's career to make the message authentic.

The goal is a higher response rate and, ultimately, more donations. You don't get points for simply sending out a mass email. If no one *acts* on the email, then it didn't work.

Remember to add details that make the person feel validated and important, and the response rate might surprise you.

Crowdfunding

HOW TO PITCH REPORTERS BEFORE, DURING AND AFTER LAUNCH

Before, during or after the launch, you'll want to contact reporters to encourage them to cover your announcement.

The email pitch changes depending on when you send it out. Below I provide both options — before and during the launch. Each one must contain a strong intro to grab the reader's attention and solid data that proves your product solves a problem.

It's also important here to think small. The *big* national media outlets are tough to reach, especially for a product no one has heard about. Do a deep dive and find niche blogs and publications in your industry. Those have smaller audiences, are easier to approach and can deliver your message to people who might care more than the general population.

Sure, a story in *The Wall Street Journal* would be huge, and if you want to go for the *big* press, I won't stop you (see my press release template on page 34). But make sure you pitch the niche blogs in your space. They matter too.

I also encourage you to email reporters individually and make a custom pitch. A mass email with everyone BCC'ed will have less of an impact.

Before the launch

Subject line: New portable fan promises to blow away the competition

Hi [reporter's first name],

I'm [first and last name], the [job title] at [name of company]. We are the makers of [name of product; for instance, "Acme Breezey"], [What's your product? Why is it special? For instance, "the first-ever portable fan powered by the sun"]. We plan to launch a campaign on [name of platform] on [month/day/year] and hope you can report on us.

NOTE: The main points are all in the opening section.

[Then, explain why you chose to contact this reporter, reference his/her work, why you like it and then link to it; for instance, "I'm reaching out because I think you'll find Acme Breezey an interesting product to cover. I <u>read your review last week</u> on the robotic pet dog, and I thought your experience with the dog at the park with other animals was hilarious."]

[Then, provide stats/metrics on the product; for instance, "Here are a few stats about Acme Breezey:

- At 18 inches, the fan is easy to carry and stow
- Has a patent-pending "Gripr" so it can attach to any surface
- Only weighs 4.5 pounds]

[Finally, give a clear call to action; for instance, "I would be happy to ship you an Acme Breezey so you can test it out for yourself."]

Thanks, and let me know if you have any questions.

– Your first name

Email signature

Deeper Insight

Keep your "before the launch" email brief but also make sure to praise the person's work and why you respect it. Make the email special as it will encourage a response.

During the launch, the only difference to the email is the intro paragraph:

I'm [first and last name], the [job title] at [name of company]. We are the makers of [name of product; for instance, "Acme Breezey"], [What's your product? Why is it special? For instance, "the first-ever portable fan powered by the sun"]. We are in the middle of our [name of platform] campaign and to date we have raised [amount] from [number of] backers. The campaign ends on [date].

After the launch, the subject line and intro paragraph change:

Subject line: New portable fan raises [amount], reaches/exceeds [fundraising platform] goal

I'm [first and last name], the [job title] at [name of company]. We are the makers of [name of product; for instance, "Acme Breezey"], [What's your product? Why is it special? For instance, "the first-ever portable fan powered by the sun"].

I'm writing to tell you we reached/exceeded our [fundraising platform] goal of [amount] with the help of [number of] backers. With the momentum, we hope you will consider our product for coverage on [name of website/ publication].

Deeper Insight

Across the three situations (before, during and after), much of the email remains the same. Why? Because in every scenario, you need to show the reporter you respect his/her work and back up your claims with hard numbers.

But it's also key to amend the intro paragraph so you explain the current situation and help the reporter understand what you want.

How to tell your audience about the launch

As you prepare for the launch of a crowdfunding campaign, it's critical to tell your own audience. Whether the message goes out through a personal email or e-newsletter (or both), you need to stay focused and include clear calls to action.

Subject line: Our [name of crowdfunding platform] campaign is live!

Hi, everyone!

We at [name of company] are excited to announce our [name of crowdfunding platform] campaign for [name of product and link to it; for instance, "Acme Breezey"] is live! Here's the campaign page.

In case you don't know, [name of product] is [one sentence about the product; for instance, "the first-ever portable fan powered by the sun"].

Our [name of crowdfunding platform] campaign will last for [number of days] days. Now we need you to spread the word and help us reach our goal of [amount]. With each new giving level, we will offer fantastic gifts and experiences. Here are a few:

[List of three levels and what the person would receive; for instance…

1. $50 — One Acme Breezey, a beer koozie and Acme Breezey t-shirt

2. $100 — Two Acme Breezeys, a beer koozie and Acme Breezey t-shirt

3. $250 — Two tickets to cocktail reception in which we publicly unveil Acme Breezey, two Acme Breezeys, two beer koozies and two Acme Breezey t-shirts

Please consider a contribution today so we can meet and exceed our goal!

How else you can help

SPREAD THE WORD! We created tweets you can use. Or <u>click on this Facebook post</u> to share it with your network.

> *NOTE: If there are other social networks you'd prefer, go with those.*

Here are the tweets:

- The @AcmeBreezey [name of crowdfunding platform] campaign is live! Let's make solar-powered portable fans a reality. Donate here: [link]

- Donate today to the @AcmeBreezey [name of crowdfunding platform] campaign! Say "yes" to solar-powered portable fans. [link]

Thanks so much for the support. Of course, reach out with any questions!

– Your name

Email signature

Deeper Insight

Once your campaign launches, give your audience two ways to help. People can either donate or encourage others to do so (via the ready-made social media posts).

It's also important to tease some of the giveaways at different giving levels within your email. Those details could entice people to click through and make a contribution.

The details ALWAYS make the difference.

Fundraising pages introduction

You have a product the world needs to see. Now, the challenge: make your "pitch" in a way that moves people to action.

What are the best writing strategies for crowdfunding campaigns? To me, an effective campaign includes three main components:

- Stories
- Data
- Clear calls to action

Stories allow you to connect with the reader on a human level. **Data** is the concrete evidence to bolster your case. And **clear calls to action** allow you to turn a passive site visitor into an engaged customer and supporter of your mission.

In the following section, I walk you through emails and content for your campaign. Each exercise is designed to make your product/service memorable and worth exploring further.

I know crowdfunding pages come in all shapes and sizes. I provide suggestions and not hard-and-fast templates.

Kickstarter project templates

Obviously, you need content on your crowdfunding page, but what should you write? How do you capture attention and convince people to support you?

I went through the project description forms on Kickstarter and Indiegogo and provided examples for several sections. They are listed on the following pages.

Yes, there are other crowdfunding sites, but these two are the most popular and many of my guides will transfer from one site to the next with slight modifications.

The fictitious company I use is called "Acme Drive Smart," which creates decals in the shape of hands that fit at the "9 and 3" areas of the steering wheel. The goal is to combat distracted driving.

Short Blurb

Example of the "short blurb" section:

Acme Drive Smart reminds you to keep your hands at "9 and 3" on the steering wheel so you're in control and not distracted by devices.

Deeper Insight

The "short blurb" must contain the value proposition: what problem do you solve? Make sure you answer that question so the person has a reason to click and learn more.

Rewards

Crowdfunding rewards should abide by two rules:

- Brief
- Descriptive

How can you be brief and descriptive at the same time? I'll show you.

Acme Drive Smart Rewards
Pledge $150 or more

Acme Prize Pack

Two sets of hand decals, bumper sticker you can customize up to 45 characters and a tote bag featuring our "hands logo" designed by famed San Francisco street artist Jan Goorman.

Brief? Check. It's a quick couple of lines. But look at the detail I managed to slip in.

- "customize up to 45 characters"
- "hands logo"
- "designed by famed San Francisco street artist Jan Goorman"

I could have written:

Two sets of hand decals, bumper sticker and tote bag.

But where's the fun in those rewards?

Project Description

The best way to make an emotional connection with the reader is to tell a story. That means you should put your product into a narrative with a beginning, middle and end. Take the reader through an actual event/experience as a way to demonstrate how your product solves a problem.

Stories are the best way to make your case (see the section on grantwriting on page 183).

Here's a fictional story for Acme Drive Smart. My recommendation is a length of 250–400 words. Not too short, not too long. It's the "right" length to tell a memorable story and keep people moving through your crowdfunding page.

Also make use of the features Kickstarter provides: bold/italics font, links and video/photos/music. Everything in moderation, but deploy the tools so you capture the reader's attention.

And don't be afraid to talk in the first person either. It's more meaningful and authentic.

Here's the example for Acme Drive Smart.

––––––

March 11, 2014. A day I will never forget.

I kicked a soccer ball with my son (five years old at the time) in the front yard of my house in Tampa, Florida. Two of my neighbor's kids (11 and 13 years old) played "HORSE" on a basketball hoop out in the street.

I saw a car turn down the road and head our way. I assumed the driver spotted the two kids playing basketball but as I looked closer, I saw the man was looking straight down at his cell phone.

Before I could even yell to the neighbor's kids, it was too late. The driver clipped one of the boys and sent him hurtling into the driveway. The other child was unhurt but stood motionless out of fear.

The child suffered a broken right leg and the lingering effects of a traumatic experience. Here's a local news article about the incident. The driver was distraught, of course, but the damage had been done — and could have been much worse.

After that day, I vowed to find a way to end distracted driving. That's why I developed Acme Drive Smart, decals in the shape of hands that fit at the "9 and 3" areas of the steering wheel. The decals are a constant reminder we should keep our eyes on the road and hands on the wheel at all times.

We need $20,000 to increase production of the decals and pay for our efforts to visit with state and national transportation officials. A huge part of our project is to build relationships with key people so Acme Drive Hands can reach motorists wherever they live.

Distracted driving needs to stop. **And with your help, it can.**

Please watch our short video below (90 seconds) and check out the awesome rewards if you contribute.

Thanks for reading, and let's make a difference for ourselves and our children!

[image of Mike's signature]

– Mike Walker

Founder and CEO, Acme Drive Smart

[Then, add in photos/videos/links below to let people understand the product. Kickstarter lets you drop in whatever details you want, but the section should begin with a compelling story as I provided above. Make the emotional connection and then back it up with hard evidence.]

Deeper Insight

The story does the selling. A compelling beginning, middle and end will captivate readers and make the case for your product.

Note the details in my story:

▸ Beginning detail: "played 'HORSE' on a basketball hoop out in the street"

▸ Middle detail: "The other child was unhurt but stood motionless out of fear"

▸ End detail: "visit with state and national transportation officials"

My specific language enriches the story and makes the product more compelling.

Risks and Challenges

Kickstarter asks, **"What are the risks and challenges that come with completing your project, and how are you qualified to overcome them?"**

You should answer the question honestly but also ensure people you have the grit and tenacity to handle any challenge.

For instance:

"Right now we have a used Acme Vinyl Cutter 1000 to create the decals in the shape of hands. We hope to use some of the money we raise to buy a second vinyl cutter and ramp up production.

Without a second machine, orders may be delayed two to three weeks so we are determined to purchase the machine ASAP and double our output."

Again, details strengthen your case. Writing "Acme Vinyl Cutter 1000" makes the team look more legit and serious than if it said something like, "a used vinyl cutter."

And after the team explains the challenge, it then writes "we are determined to purchase the machine ASAP." That's strong, confident language which will hopefully allay any concerns from potential backers.

About You

In the "About You" section, Kickstarter allows you to provide a biography. Again, I return to the importance of a bio that lets people understand who you are and what you're passionate about. I first provided the template in the LinkedIn section on page 123.

Now I will apply the same concept to the Kickstarter page. In total, the bio takes about 30 seconds to read and, like on LinkedIn, makes a human, emotional connection with the reader.

Step 1: Who are you, really?

Acme Drive Smart is a team committed to eliminating distracted driving from our roadways.

NOTE: In one clear line, this is who we are and what we do.

Step 2: What do you do and what makes you qualified for someone's investment?

At Acme, we have spent 15 months testing and developing the most effective decals in the shape of hands that wrap around the "9 and 3" areas on a steering wheel. Through research and development, we now have a product that will make driving safer for everyone. Our product won first prize at the "North Carolina StartUp Challenge" and gave us added confidence we are on the right track.

> NOTE: Deeper explanation with details like expertise (ex: 15 months testing and developing) and honors/distinctions if available (ex: "first prize" at the startup challenge). You could also list off different media outlets that have covered your product.

Step 3: Bring 'em home

Distracted driving needs to stop, and we feel Acme Drive Smart is the product to end it for good. To learn more, visit our website below!

> NOTE: Closing line that sums up why you're passionate about the work and helping others.

Indiegogo project description templates

Campaign tagline

For the campaign tagline, treat it like a headline to an article or blog post:

> Eliminate distracted driving with
> Acme Drive Smart decals

Short and sweet with a value proposition ("eliminate distracted driving").

Campaign Overview

Indiegogo asks that you **"Lead with a compelling statement that describes your campaign and why it's important to you."**

Here's an example for Acme Drive Smart, and remember it should flow from the tagline right above it.

Eliminate distracted driving with
Acme Drive Smart

Acme Drive Hands are sharp, colorful decals in the shape of hands that remind you to hold the wheel at "9 and 3." It's the smartest, safest way to drive any car night and day.

———

The tagline and overview set the tone and let readers know what you're about — and why it should matter to them.

Campaign Pitch

Indiegogo lays out a formula for the "Campaign pitch" area. It is:

- Short Summary
- What We Need & What You Get
- The Impact
- Risks & Challenges
- Other Ways You Can Help

I don't recommend a giant blob of text, though. Do your best to intersperse photos, graphics, video and other eye-catching elements to keep the reader engaged.

But when it comes to the content itself, here's what I recommend. Again, the product is Acme Drive Hands.

Short Summary

NOTE: I abbreviated the story from Kickstarter to fit in the flow of the content on Indiegogo. I still make the story personal (it's in the first person) to connect with the reader.

March 11, 2014. A day I will never forget.

As I played soccer with my son (five years old at the time) in front of my house in Tampa, I watched as a driver ran into one of my neighbor's kids as he played basketball out the street.

The cause: distracted driving.

The child suffered a broken right leg and the lingering effects of a traumatic experience. Here's a local news article about the incident. The driver was distraught, of course, but the damage had been done — and could have been much worse.

After that day, I vowed to find a way to end distracted driving. That's why I developed Acme Drive Smart, decals in the shape of hands that fit at the "9 and 3" areas of the steering wheel. The decals are a constant reminder we should keep our eyes on the road and hands on the wheel at all times.

With your support, my team and I can produce the decals in greater quantities and visit with state and national transportation officials to share product samples.

Distracted driving needs to stop. **And with your help, it can.**

What We Need & What You Get

We need $20,000 to increase production of the decals and pay for our efforts to meet with various people in the transportation community. A huge part of our project is to build relationships with key people so Acme Drive Hands can reach motorists wherever they live.

Example of a perk

Pledge $150
Acme Prize Pack
Two sets of hand decals, bumper sticker you can customize up to 45 characters and tote bag featuring our "hands logo" designed by famed San Francisco street artist Jan Goorman.

> *Note: (Visit page 199 to learn why details make the difference in the rewards area)*

The Impact

With your help, we believe we can make strides in the battle against distracted driving. Too many of us look away from the wheel to check our phones, reapply make up or some other unnecessary task. With Acme Drive Smart, we give drivers a constant reminder to place their hands in the right spots and keep their eyes on the road.

Risks & Challenges

Right now we have a used Acme Vinyl Cutter 1000 to create the decals in the shape of hands. We hope to use some of the money we raise to buy a second vinyl cutter and ramp up production.

Without a second machine, orders may be delayed two to three weeks so we are determined to purchase the machine ASAP and double our output.

Other Ways You Can Help

Spread the word about Acme Drive Hands on your favorite social media channel! Tell your network you want to solve the societal problem of distracted driving. Make sure to use the social buttons at the top of the page!

Handwritten thank-you note to top backers

For people who donated at high levels, you should consider handwritten thank-you notes. It's a "next-level" move to show you appreciate the person, and it helps to build trust.

You should also make the message personal to whatever degree possible. Since you might not know the person, play off of where he/she lives.

Here's an example for Acme Drive Smart.

Hi [backer's first name],

Thank you so much for supporting the crowdfunding campaign for Acme Drive Smart. Your donation will help us make the roads safer all across the nation and (hopefully one day) the world. I will be sure to keep you posted on our progress. Have a great summer down in sunny San Diego — it's on my list of places I need to visit!

Thanks again,

– Your first name

___

Deeper Insight

Did you catch the personalized section? The part about visiting "sunny San Diego" shows I didn't write the exact same thank-you note 50 times.

Above all, a handwritten thank-you note is a classy move. Again, if you have hundreds of backers at lower levels, a standard email thank-you note is enough. But for those who went above and beyond, you need to respond accordingly.

My Journey: Part 7
That time I realized I have so much left to learn

I walked into McCormick Place in Chicago with wide eyes and my mouth agape.

I knew BookExpoAmerica (BEA) was *the* event for the publishing world, but I was blown away by the scene in front of me.

Every major publisher had a massive "booth" with book shelves, couches, tables and checkout counters. It felt like a giant mall of bookstores one after the next spread across a room the size of three football fields.

I wandered through a big row of "booths" for ten minutes and realized right away I had a lot to learn about the book biz.

My day at BEA kickstarted an education into not only book publishing but also what it takes to run a business (traditional sales and online).

I understood it's not enough to "write a book." There's *so much more* to the process, especially if I wanted to operate as an independent author.

Writing, publishing, selling — the onus is on me to make everything hum. And that means the learning never stops. If I don't "figure it out," who will?

So the education goes on even today. And like anyone else determined to pursue their passion, the knowledge is essential to move from one stage to the next.

And with each new step, the story — *my* story — becomes more interesting.

Chapter 8
Public Speaking

WHY YOU NEED TO CRITIQUE YOUR OWN VOICE

For many people, listening to their own voice makes them feel awkward. They often say, "Ugh, I can't believe I sound like that!"

Unless you work in broadcasting or regularly deliver speeches, you rarely hear yourself in a professional context.

I know it's painful, but you need to hear yourself talk. As I promote my business, I have opportunities to do podcasts, radio interviews and in-person appearances. Each time, I ask for the recording (and, when available, video) so I can listen/watch all the way through.

Why listen to myself? Several reasons. I want to see if:

▸ My arguments were solid and logical

▸ I stutter or say "um," "like" and "you know"

▸ I talk too fast or rush a particular answer

▸ I handle every question the right way

▸ I have enough energy in my voice

▸ I move in a strange way or have an odd mannerism

I'm determined to make each speaking opportunity better than the last and the only way is to (gulp) scrutinize every performance.

What you can do

The next time you prepare for a speech or presentation, record the audio (and maybe video too) at your desk. Stand up when you read because it gives your voice more energy.

Then, press play and listen.

Ask yourself:

- Do I need more pep in my voice?

- Do I go too fast or too slow?

- Do I put emphasis on the right words?

- Am I bored by my own voice? (Seriously, would *you* listen to you?)

How we sound to others makes a huge difference in our careers. Job interviews, networking, phone calls. All the time.

Remember: confidence is king.

Two words every speaker should avoid

In this section, I focus on a widespread public speaking hiccup: the use of the words "uh" and "um."

Yes, we're all guilty of the two verbal crutches, but some speakers lean on them way too much. After a few minutes, the audience starts to wait for the next "uh" or "um" and may not process what the person has to say.

If we've gone years sprinkling "uh" and "um" into daily conversations, then the habit is magnified when we're nervous in front of a crowd.

So what to do? How do we break away from "uh" and "um"? For me, the answer is...

...did you catch it?

That's right. *Silence.* If I don't have the right word or am searching for the next sentence, I don't say anything. I pause, stay quiet and plan my move.

I'm not silent for an extended length of time like 10 seconds. That would be weird. It's more like one to two seconds, those little bits throughout a speech where I could drop an "uhhhh" and fill the gap. Instead, *nothing.*

The approach has another, more subtle advantage too — it helps to draw the audience in. If I create brief moments of quiet, it makes people lean closer and think, "What's he going to say next?"

On the flipside, if I talk in a never-ending stream of sentences linked by "uhs" and "ums," there's no time for the audience to absorb the thought. It's like a moving train they can't board. **Frustrating.**

With little pockets of silence, everyone (speaker too) has time to think about the most recent line and gear up for the next one.

When you talk to a group (even from your seat at a staff meeting), try to pause rather than cling to "uh" and "um."

Sure, the words will keep you afloat, but your message may be lost at sea.

How to introduce yourself on the phone

When you make a work-related phone call, what's your move after the person says, "Hello?"

Too often, I hear people do the following:

"Hi, is ____ there?"

To which the person is forced to say:

"Can I ask who's calling?"

Of course the person needs to ask who's on the line. We never said our name — how would he/she ever know?

That's why, in only three seconds, you can impress people on the phone with a simple strategy: introduce yourself right away.

"Hello?"

"Hi, my name is Jane Doe from Acme Industries. Is John there?"

"Sure, let me get John for you."

"Great. Thanks."

See the difference? Plus, we sound much more confident if we lead with our first and last name and *then* ask for the person. Yes, it's a subtle move and takes up one percent of the phone call. Still, a proper introduction sets the tone.

Remember, in the business world you need to impress everyone at every turn. Let's say you call a company to promote your services. Each time, a receptionist answers with "Acme Industries. How can I help you?"

You respond with:

"Hi, my name is Jane Doe, and I'm calling to speak with John about my business, Tech Corporation."

"OK sure. Let me connect you to John's office."

A proper introduction is a small detail, but maybe you score points with the secretary. Maybe the boss wants the secretary's opinion on who to do business with. Maybe your phone etiquette helps you land the business. Ya never know.

What I *do* know is when "Hello?" happens, you need to be ready. It's a little word with big opportunity.

How to leave a proper voice mail

I get it. If I call a friend and he doesn't answer, I hang up and send a text.

Other times, it's not so easy. If I call someone for business, the hang-up-and-text method doesn't work. I need to leave a professional voice mail and say all the right stuff as soon as the **BEEEEEP** comes in.

I know career-related phone calls can be awkward and even nerve wracking. Here are four guidelines to make your voice mails polished and poised.

1. Don't go on forever

Isn't it the worst when someone holds an entire (one-sided) conversation in a voice mail? Explain who you are and briefly why you're calling. Then say goodbye and hang up.

2. Always start with your name (Hi, my name is _____)

Don't make people guess the voice in the voice mail. By the time they figure it out, they didn't pay attention to anything you said.

3. Give contact info S-L-O-W-L-Y

It's annoying when someone says a phone number or email address quickly — and only one time — and then hangs up. Then I need to play the message a second time to write down the number.

Say your number slowly, digit by digit, and then do it a second time. Same goes for an email address. Letter by letter.

4. Repeat your name at the end

You know when you meet someone and never catch the name the first time around? You shake hands, say hello but then think, "Wait, what was her name again?" Same for the voice mail. Repeat your name and company at the end so the person is 100% clear on who left the message.

If you follow steps 1–4, you will leave perfect voice mails every time.

My Journey: Part 8
Conclusion: That time I understood the power of storytelling

You can TELL people your company solves a problem and hope the audience believes you.

Or you can SHOW your value through a compelling story of success.

In *Wait, How Do I Promote My Business?*, I've taken you on a journey (AKA story) through my early experience as an author, publisher and entrepreneur. In each chapter of the story, I also demonstrated how I used templates from my book to overcome obstacles and create new opportunities.

Why? To prove that my writing guides work. And to underscore the raw power of a memorable story.

When you need to convince skeptical outsiders that your company/organization delivers value, share customer or client success.

The anecdote adds a human element to an otherwise abstract product or service. It also allows other people to imagine themselves benefiting from your offer.

213

And no matter the writing challenge — website content, press release, email outreach and the list goes on — remember to speak from the heart with a blend of authenticity, curiosity and storytelling. These three qualities are the hallmarks of any effective communicator and the central themes of my book.

May this book — combined with your hard work — help your business reach new heights.

Then *you* can write a book of your own with a fitting first chapter:

That time my writing skills helped me open new doors.

Chapter 9
Thank-Yous

If you read my author journey throughout the book ("That time I…"), then you know it was a major task to assemble my first book, *Wait, How Do I Write This Mail?*

I didn't know how to start or who would guide me. Total blank slate.

The second time around? Way less stressful.

I returned to the same people who helped me the first time. So thank you — again — to Paul McCarthy (paulmccarthydesign.com) for the front cover design and a big thanks to 1106 Design (1106design.com) for the book's interior layout.

Wow, it was so much easier to construct *Wait, How Do I Promote My Business?* because I had — ironically — a template in place from book #1.

Thank you to my patient wife, Shikma, who allowed me to work on this book at odd hours and on weekends. You know it's my passion to teach business communication skills. And you forever support my decision to not only write books but also speak on the topic (sometimes out on the road).

A big shout out to the group of people who served as my focus group and helped me build out my list of challenging writing scenarios. Many thanks to Erik Olson, Kaushik Saha, J.J. Peller, Andre Teow, Danielle Benson and Matt Putterman. Your insights expanded my thinking and, ultimately, the table of contents.

Thanks to the countless people who sent me template suggestions via email or my blog. Like my first book, *Wait, How Do I Promote My Business?* is a crowd-sourced effort focused on our collective writing challenges.

And thanks to my many public relations clients over the years who hired me to help with their own content. With each project, I added to my list of writing challenges most business owners face. I always thought, "Well, if one business owner needed the assistance, maybe others do too?"

Thus, *Wait, How Do I Promote My Business?* became a reality.

Yes, I wrote the book "alone," but many people left their imprint on the finished product. Whether you supported me for a minute, a month or more, thank you for your time and assistance.

The book wouldn't be the same without you.

Chapter 10
Classroom Activities & Free Webinar

I want *Wait, How Do I Promote My Business?* to help people both in the business world and the classroom.

Whether it's a business school, entrepreneurship class or some other program that preps people for the "real world," let's learn practical writing skills now and be more competitive right out of the gate.

Below I have provided three different classroom activities that draw upon lessons and templates from the book. That means students will need to use the book to complete the activities.

Each exercise contains:

▸ activity for students

▸ teacher notes

▸ estimated class time

Please visit DannyHRubin.com/teaching as I often add new activities to the mix.

Classroom activity sample

HOW TO WRITE A PRESS RELEASE

In the activity below, you will learn to write your own press release. Go through each part of the release carefully and be ready to share your work with the class.

Instructions:

- ▸ Review the press release "General Outline" section on page 34 to understand the foundation of a solid release.
- ▸ Next, write the "Header" information as described on page 35. If it's a mock exercise, you can make up the name of the company, company address and the person who is the media contact.
- ▸ Then, compose the "Headline/Subheadline" section of the press release (page 36).
- ▸ Next, work on the body of the press release (page 37).
- ▸ Finally, review all the pieces of the press release at one time.
- ▸ Be ready to share your press release and how you attempt to grab the media's attention.

Teacher notes: How to write a press release
Estimated class time: 1–2 hours

The impact
Wow, what an opportunity — to learn how to write a winning press release while a student.

In the activity, your students will craft a press release about their new product, company or upcoming event. You can then decide if you want your students to send the press release to the media (if it's a real announcement) or keep it in-house as a classroom exercise.

Skills the students learn:

- ▸ Clear communication
- ▸ Importance of using strong detail

- Respect for a person's time
- How to do basic research

Notes for the exercise:

- Consider stopping after the students write each section of the press release so they can share their work with the group.
- Spend the most time on the body of the press release. It might even span two class sessions (and the students can finish the release as homework).
- Tell your students to read the press release out loud (quietly at their seats) to make sure there are no typos or awkward phrases.

Questions to ask the students:

- Why did you choose your particular headline/subheadline?
- Who can provide an example of a number or data point they used in their press release? How does the number enhance your announcement?
- What is your call to action for the media?

Classroom activity sample

HOW TO INTRODUCE YOURSELF TO A COMPANY FOR THE FIRST TIME

In the following activity, you will learn to build a relationship with a potential business partner/connection the right way. It's all about authenticity and curiosity. Take your time constructing the email and be ready to share your work with the class.

Instructions:

- Review the introductory email guide on page 78 to understand the foundation of a solid outreach message.
- Before you begin to write, research the company/organization and find at least two recent projects/initiatives that stand out to you (example from page 78 is "Bark Bark 5K Race"). Also prepare at least two details on your own business

that the email recipient would find relevant and notable (example from page 79 is "4,500 units").

▶ With your research completed, draft the entire email. Be sure to put the description in your own words and don't copy language from the example in the book.

▶ Finally, review the email by reading the message out loud (quietly) to check for style and grammar.

▶ Be ready to share your email and how you attempt to earn the trust and respect of the recipient.

Teacher notes: How to introduce yourself to a company for the first time
Estimated class time: 30–40 minutes

The impact

In business, relationships are everything. And the email exercise here contains the basics of how to gain a person's trust and respect in a "cold" email — a message the person is inclined to ignore.

And I would encourage your students to send these emails to the intended recipient rather than treat it as a practice run. Help them make the lesson real!

Skills the students learn:

▶ Curiosity in others

▶ Importance of using strong detail

▶ Respect for a person's time

▶ How to do basic research

▶ Relationship building

▶ Advanced business networking

Notes for the exercise:

▶ The research component of the activity is critical. Your student is not allowed to write, "I think your company does great work." He/she needs to select specific examples from the company website of projects they respect. That way, the email will not feel "canned" and as though the student sent the same message to 50 companies.

▸ Encourage students to read their finished emails out loud so others can hear what they did.

Questions to ask the students:

▸ What research did you choose to include on the company and why?

▸ What examples of your own company success did you offer and why?

▸ Why do you think it's important to tell someone else *specifically* why you respect the work they do?

_____

Classroom activity sample

HOW TO WRITE A CROWDFUNDING PROJECT DESCRIPTION

In this sample activity, you will learn to compose a project description for a crowd-funding campaign — or perhaps a campaign you might stage at a later date. **Remember** the best way to convince someone is to tell a memorable story about how your product/service originated.

Instructions:

▸ Review the Kickstarter "Project Description" on page 199 to see an example of an effective description.

▸ Brainstorm the story you want to share. Often it's the story behind the reason for the product or service you created. Explain the inspiration behind the project. What moment or experience spurred the idea?

▸ Next, write the description by following the model on page 199. Be sure to put the description in your own words and don't copy language from the example in the book.

Teacher notes: How to write a crowdfunding project description
Estimated class time: 45 minutes to 1 hour

The impact

The exercise teaches students the power of storytelling as a way to move people to action. A story captures readers' emotion and encourages them to, in this case, donate to a project.

Skills the students learn:

► Clear communication

► Importance of using strong detail

► Storytelling

► Respect for the reader's time

Notes for the exercise:

► Do the story brainstorm session as a group. Encourage the students to think about the idea/moment that led to the creation of their product or service. What experience prompted the idea? That can often be a great place to locate the story.

► Make sure the students' stories have a beginning, middle and end. If students leave the story in the middle, then the reader needs to ask, "Well, how did it end?"

► The description needs to be in the first person. Don't let the students write in the third person ("At Acme Industries, we…"). First person is the most authentic approach.

Questions to ask the students:

► What story did you choose and why?

► How did you transition from the story to your campaign goal?

► Which words did you put in bold or italics and why?

► Why do you think storytelling is an effective approach on a crowdfunding page?

Register for Danny's Free Webinar!

For deeper instruction, register now for Danny's FREE webinar on how to write stellar job applications.

In the webinar, Danny will change your thinking on every major aspect of the job application process. By the end, you won't look at your resume/cover letter the same way again. You will learn to:

▸ Construct your resume so your skills and experience **JUMP** off the page

▸ Transform your cover letter into a **powerful** short story of success—a technique that can grab the employer's attention immediately

▸ Craft your LinkedIn profile summary so you make the **strongest** "introduction" before the employer ever contacts you

▸ Prepare for the job interview with strategies that make you **unfor-gettable** (most applicants **NEVER** consider them at all!)

Visit DannyHRubin.com/jobappwebinar to claim your spot!

Chapter 11
Need Professional PR Help?

My thanks to Renee Wilson, president of the PR Council, for contributing the checklist below. PR Council is the national association of PR agencies, and Wilson provides expert advice on how to select a PR firm that's best for your business.

Wilson wrote me: "Client-agency partnerships are like any other relationship—you will get out of it, what you put in. That is why it's important to take the time to find the right partner."

TOP 10 CRITERIA TO CONSIDER

1. Industry-specific expertise: If your brief is about launching a product in a specific industry such as food, automotive or biotech, you may want to consider a firm that specializes in that area.

2. Budget allowance: Carefully consider the amount of budget that you want to put against your public relations efforts as different size firms have different budget requirements. It doesn't make one firm better or worse, it's just reflective of the size or type of assignment.

3. Your available resource: Determine who from your organization will be managing your new PR agency. If your resource is limited, and quite junior in experience, you may want an agency that can truly take the brief and run with it. However, if you plan on being quite hands on, your agency selection may vary.

4. Target audience: It's important to think through the audience(s) you wish to target as certain firms specialize in various audiences. For example, if you had an assignment that was about "marketing to moms," there are firms that specialize in that area. If you had an assignment that required targeting financial investors, other firms will specialize in that.

5. Cultural fit: The culture of a firm will say a lot about its values, behaviors and the company it keeps. Be sure to spend enough time 'dating' your prospective agency to ensure your corporate cultures are aligned.

6. Geography desired: Think about whether or not you need local, national or international reach and expertise. This will help determine the type of agency you need and where they are located.

7. Senior-level involvement: Do you need and/or want senior level counselors regularly engaged in your business? If so, be sure to select an agency where that is a part of your agreement.

8. Team dynamics: Look at the chemistry of the group of people who will make up your team. Are they a unit you can see yourself working with in a productive manner, day after day?

9. Project vs Agency of Record (AOR): If your assignment is more project-oriented than long-term engagement, you may want a certain type of firm. Perhaps you want a smaller firm to work on a smaller project? Alternatively, you may want a small team within a larger firm but the point is: think about how a project versus AOR will work for the engagement.

10. Traditional vs. integrated capability: Most PR agencies these days offer some form of integrated capabilities (ex: assist with advertising efforts in addition to PR services). However, not all do. If you are not looking for all of that, and only need media relations work done, you may opt for a traditional firm, with a very focused skill set.

Chapter 12
Index